Titles include:

Palgrave Studies in Nineteenth-Century Writing and Culture
Series Standing Order ISBN 978–0–333–97700–2 (hardback)
(*outside North America only*)

You can receive future titles in this series as they are published by placing a standing order. Please contact your bookseller or, in case of difficulty, write to us at the address below with your name and address, the title of the series and the ISBN quoted above.

Customer Services Department, Macmillan Distribution Ltd, Houndmills, Basingstoke, Hampshire RG21 6XS, England

Palgrave Studies in Nineteenth-Century Writing and Culture

General Editor: **Joseph Bristow**, Professor of English, UCLA

Editorial Advisory Board: **Hilary Fraser**, Birkbeck College, University of London; **Josephine McDonagh**, Kings College, London; **Yopie Prins**, University of Michigan; **Lindsay Smith**, University of Sussex; **Margaret D. Stetz**, University of Delaware; **Jenny Bourne Taylor**, University of Sussex

Palgrave Studies in Nineteenth-Century Writing and Culture is a new monograph series that aims to represent the most innovative research on literary works that were produced in the English-speaking world from the time of the Napoleonic Wars to the *fin de siècle*. Attentive to the historical continuities between 'Romantic' and 'Victorian', the series will feature studies that help scholarship to reassess the meaning of these terms during a century marked by diverse cultural, literary, and political movements. The main aim of the series is to look at the increasing influence of types of historicism on our understanding of literary forms and genres. It reflects the shift from critical theory to cultural history that has affected not only the period 1800–1900 but also every field within the discipline of English literature. All titles in the series seek to offer fresh critical perspectives and challenging readings of both canonical and non-canonical writings of this era.

Titles include:

Eitan Bar-Yosef and Nadia Valman (*editors*)
'THE JEW' IN LATE-VICTORIAN AND EDWARDIAN CULTURE
Between the East End and East Africa

Heike Bauer
ENGLISH LITERARY SEXOLOGY
Translations of Inversions, 1860–1930

Laurel Brake and Julie F. Codell (*editors*)
ENCOUNTERS IN THE VICTORIAN PRESS
Editors, Authors, Readers

Luisa Calè and Patrizia Di Bello (*editors*)
ILLUSTRATIONS, OPTICS AND OBJECTS IN NINETEENTH-CENTURY LITERARY AND VISUAL CULTURES

Deirdre Coleman and Hilary Fraser (*editors*)
MINDS, BODIES, MACHINES, 1770–1930

Colette Colligan
THE TRAFFIC IN OBSCENITY FROM BYRON TO BEARDSLEY
Sexuality and Exoticism in Nineteenth-Century Print Culture

Eleanor Courtemanche
THE 'INVISIBLE HAND' AND BRITISH FICTION 1818–1860
Adam Smith, Political Economy, and the Genre of Realism

Dennis Denisoff
SEXUAL VISUALITY FROM LITERATURE TO FILM, 1850–1950

Stefano Evangelista
BRITISH AESTHETICISM AND ANCIENT GREECE
Hellenism, Reception, Gods in Exile

Margot Finn, Michael Lobban and Jenny Bourne Taylor (*editors*)
LEGITIMACY AND ILLEGITIMACY IN NINETEENTH-CENTURY LAW, LITERATURE AND HISTORY

Yvonne Ivory
THE HOMOSEXUAL REVIVAL OF RENAISSANCE STYLE, 1850–1930

Colin Jones, Josephine McDonagh and Jon Mee (*editors*)
CHARLES DICKENS, A TALE OF TWO CITIES AND THE FRENCH REVOLUTION

Jarlath Killeen
THE FAITHS OF OSCAR WILDE
Catholicism, Folklore and Ireland

Stephanie Kuduk Weiner
REPUBLICAN POLITICS AND ENGLISH POETRY, 1789–1874

Kirsten MacLeod
FICTIONS OF BRITISH DECADENCE
High Art, Popular Writing and the *Fin de Siècle*

Diana Maltz
BRITISH AESTHETICISM AND THE URBAN WORKING CLASSES, 1870–1900

Catherine Maxwell and Patricia Pulham (*editors*)
VERNON LEE
Decadence, Ethics, Aesthetics

Muireann O'Cinneide
ARISTOCRATIC WOMEN AND THE LITERARY NATION, 1832–1867

David Payne
THE REENCHANTMENT OF NINETEENTH-CENTURY FICTION
Dickens, Thackeray, George Eliot and Serialization

Julia Reid
ROBERT LOUIS STEVENSON, SCIENCE, AND THE *FIN DE SIÈCLE*

Virginia Richter
LITERATURE AFTER DARWIN
Human Beasts in Western Fiction 1859–1939

Anne Stiles (*editor*)
NEUROLOGY AND LITERATURE, 1860–1920

Caroline Sumpter
THE VICTORIAN PRESS AND THE FAIRY TALE

Sara Thornton
ADVERTISING, SUBJECTIVITY AND THE NINETEENTH-CENTURY NOVEL
Dickens, Balzac and the Language of the Walls

Ana Parejo Vadillo
WOMEN POETS AND URBAN AESTHETICISM
Passengers of Modernity

Phyllis Weliver
THE MUSICAL CROWD IN ENGLISH FICTION, 1840–1910
Class, Culture and Nation

Paul Young
GLOBALIZATION AND THE GREAT EXHIBITION
The Victorian New World Order

Palgrave Studies in Nineteenth-Century Writing and Culture
Series Standing Order ISBN 978–0–333–97700–2 (hardback)
(*outside North America only*)

You can receive future titles in this series as they are published by placing a stand-ing order. Please contact your bookseller or, in case of difficulty, write to us at the address below with your name and address, the title of the series and the ISBN quoted above.

Customer Services Department, Macmillan Distribution Ltd, Houndmills, Basingstoke, Hampshire RG21 6XS, England

---

Minds, Bodies, Machines, 1770–1930

Edited by

Deirdre Coleman
Robert Wallace Chair of English, University of Melbourne, Australia

and

Hilary Fraser
Geoffrey Tillotson Professor in Nineteenth-Century Studies and Executive Dean of the School of Arts, Birkbeck, University of London

palgrave
macmillan

First published 2011 by
PALGRAVE MACMILLAN

Palgrave Macmillan in the UK is an imprint of Macmillan Publishers Limited,
registered in England, company number 785998, of Houndmills, Basingstoke,
Hampshire RG21 6XS.

Palgrave Macmillan in the US is a division of St Martin's Press LLC,
175 Fifth Avenue, New York, NY 10010.

Palgrave Macmillan is the global academic imprint of the above companies
and has companies and representatives throughout the world.

Palgrave® and Macmillan® are registered trademarks in the United States,
the United Kingdom, Europe and other countries.

ISBN 978–0–230–28467–8   hardback

A catalogue record for this book is available from the British Library.

Library of Congress Cataloging-in-Publication Data

Minds, bodies, machines, 1770–1930 / [edited by] Deirdre Coleman,
   Hilary Fraser.
      p. cm.
   Includes index.
   ISBN 978–0–230–28467–8 (hardback)
   1. English language—Social aspects—History—19th century.
      2. Technological innovations—Social aspects—History—19th century.
      3. Literature and technology—History—19th century.   I. Coleman,
      Deirdre.   II. Fraser, Hilary, 1953–
   PE1074.75.M56 2011
   809'.93356—dc22   2011004892

10   9   8   7   6   5   4   3   2   1
20 19 18 17 16 15 14 13 12 11

Contents

Illustrations

Acknowledgements

The starting point for this collection of essays was an Australian Research Council (ARC) Linkage grant between the English and Theatre Studies programme, University of Melbourne, and software developers Constraint Technologies International (CTI). The editors would like to thank the ARC and CTI, in particular our collaborator Paul Hyland, for supporting an entire programme of activities around the subject of 'Minds, Bodies, Machines', including an interdisciplinary conference convened in 2007 by the Centre for Nineteenth-Century Studies, Birkbeck, University of London. We also thank Birkbeck and the British Academy for their support of this conference, material from which provided the focus in 2008 for a special issue of the journal *Nineteen: Interdisciplinary Studies in the Long Nineteenth Century* (www.19.bbk.ac.uk).

The aim of the conference was to explore the relationship between minds, bodies and machines in the long nineteenth century, with a view to understanding the history of our technology-driven, post-human visions, and we are grateful to all who contributed to an immensely exciting intellectual event. The editors would especially also like to thank our postgraduate students for their administrative and editorial support throughout the project, in particular, Katherine Inglis, Fiona Brideoake, Amelia Scurry and Sashi Nair. Funding for the book's illustrations was provided by the School of Culture and Communication and the Faculty of Arts, University of Melbourne.

Finally, the editors would like to acknowledge Reaktion for their kind permission to reprint Steven Connor's chapter 'Air-Looms and Influencing Machines' from *The Matter of Air: Science and Art of the Ethereal* (2010).

Every effort has been made to contact the copyright holders, but if any have been inadvertently overlooked the publishers will be pleased to make the necessary arrangements at the first opportunity.

Deirdre Coleman
Hilary Fraser

Notes on Contributors

Marie Banfield is a PhD student at Birkbeck, University of London, at present completing a thesis on the novels and poems of George Meredith in their contemporary intellectual and scientific context. Her article 'Meredith, Mazzini and the Risorgimento: The Italian Novels', was published in *Rivista di Studi Vittoriani*, XI, 21 (January 2006). Her essay 'Meredith's Altruist: Evolution, Society and Politics in *Beauchamp's Career*' forms a chapter in *Beauchamp's Career. George Meredith: testo e contesto*, a cura di Anna Enrichetta Soccio (Roma: Aracne, 2008). Her article 'Darwinism, Doxology and Energy Physics: The New Sciences and the Poetry and Poetics of Gerard Manley Hopkins' appeared in *Victorian Poetry*, 45, summer 2007.

Daniel Brown is a Professor in the English Department at the University of Western Australia. He has written books and journal articles on nineteenth-century science, philosophy and poetry; Victorian non-fiction; Gerard Manley Hopkins, Thomas De Quincey; film; and George Egerton. He recently finished writing a study of poetry by Victorian scientists, entitled *Science and Nonsense: Reading Victorian Scientists through Their Poetry*.

Deirdre Coleman's research interests include eighteenth and nineteenth-century literature, science and cultural history; abolitionism, travel writing, colonialism, natural history and racial ideology. Her articles have been published in *ELH*, *Eighteenth-Century Life* and *Eighteenth-Century Studies*, and her most recent books are *Romantic Colonization and British Anti-Slavery* (Cambridge University Press, 2005), and the *Australia* volume of *Women Writing Home, 1700–1920: Female Correspondence across the British Empire*, 6 vols (Pickering and Chatto, 2006). She is the Robert Wallace Chair of English at the University of Melbourne.

Steven Connor is Professor of Modern Literature and Theory at Birkbeck, London and Academic Director of the London Consortium Graduate Programme. He is the author of books on Dickens, Beckett, Joyce and other writers, as well as works of literary and cultural theory such as *Postmodernist Culture* (1989) and *Theory and Cultural Value* (1992). More recent books include *Dumbstruck: A Cultural History of Ventriloquism* (2000), *The Book of Skin* (2003), *Fly* (2006) and *The Matter of Air* (2010).

His website, www.stevenconnor.com, gives access to many unpublished essays, broadcasts and lectures.

Paul Crosthwaite is a Lecturer in English Literature and a member of the Centre for Critical and Cultural Theory at Cardiff University. His publications include *Trauma, Postmodernism and the Aftermath of World War II* (Palgrave Macmillan 2009), articles in *Angelaki, Cultural Politics* and *Textual Practice* and, as editor, *Criticism, Crisis and Contemporary Narrative: Textual Horizons in an Age of Global Risk* (Routledge, 2011). He is currently writing *Reading the Markets: Finance, Feeling and Representation in Contemporary Literature and Culture* and developing a new project, provisionally entitled *'An End and A Beginning': The Great Crash of 1929 and the Poetics of History.*

Hilary Fraser holds the Geoffrey Tillotson Chair of Nineteenth-Century Studies at Birkbeck, University of London, where she is also Dean of Arts. She has written monographs on aesthetics and religion in Victorian writing, the Victorians and Renaissance Italy, nineteenth-century non-fiction prose, and gender and the Victorian periodical. She currently works on women writing about art in the nineteenth century.

Katherine Inglis received her PhD in 2009 from Birkbeck, University of London. Her primary research interests are nineteenth-century fiction, the intersections between literature and science, and the history of book destruction. Her thesis, 'The Incoherent Self and Materiality in Hogg, Brontë and Dickens', explores how incoherence in fiction is supplemented by the eloquence of the material culture of science. Her publications include '"My ingenious answer to that most exquisite question": Perception and Testimony in Winter Evening Tales', *Studies in Hogg and his World*, 17 (2006); 'Becoming Automatous: Automata in *The Old Curiosity Shop* and *Our Mutual Friend*', *19: Interdisciplinary Studies in the Long Nineteenth Century*, 6 (2008), at: www.19.bbk.ac.uk; and 'Ophthalmoscopy in Villette', *Journal of Victorian Culture,* 15.3 (2010), 348–69.

Iain McCalman was born in Nyasaland, Africa and now lives in Sydney, where he is a Research Professor at the University of Sydney. He has held numerous visiting research fellowships in Britain and the United States, most recently at Duke University, North Carolina. In 2007 he was made an Officer of the Order of Australia for services to history and humanities. His most recent book, *Darwin's Armada*, has been published in Australia, the USA and the UK and was made into a three-part documentary, *Darwin's Brave New World* (2009).

James Mussell is Lecturer in English at the University of Birmingham. He is the author of *Science, Time and Space in the Late Nineteenth-Century Periodical Press* (Ashgate, 2007) and one of the editors of the *Nineteenth-Century Serials Edition* (2008: www.ncse.ac.uk). He has published widely on nineteenth-century science, print and culture. He also works in the digital humanities and is the editor of the 'Digital Forum' of the *Journal of Victorian Culture*.

Peter Otto is Professor of English Literary Studies at the University of Melbourne, where he teaches and researches in the literatures and cultures of modernity, from Romanticism to the new media of today. His recent publications include *Blake's Critique of Transcendence* (Oxford University Press, 2000); *Gothic Fiction, a Microfilm Collection of Gothic Novels Published between 1764 and 1830* (Adam Matthews, 2002–3); *Gothic Fiction: A Guide* (Adam Matthews, 2004) and *Entertaining the Supernatural: Mesmerism, Spiritualism, Secular Magic and Psychic Science* (Adam Matthews, 2007–8). His new monograph, *Multiplying Worlds: Romanticism, Modernity and the Emergence of Virtual Reality* (Oxford University Press, 2011), uncovers a key stage in the development of modern discourses of virtual reality.

Laura Salisbury is RCUK Fellow in Science, Technology and Culture and a lecturer at Birkbeck, University of London. She has published widely on Beckett, including a study of his 'aphasic' modernism, and on relations between literature and philosophy. With Andrew Shail she is co-editor of *Neurology and Modernity: A Cultural History of Nervous Systems, 1800–1950* (Palgrave Macmillan, 2010).

Introduction

Deirdre Coleman and Hilary Fraser

Which of us has not stood, captivated, before one of the 'human statues' that are now such a feature of tourist sites the world over? In October 2006, so many such figures lined the famous boulevard Las Ramblas in Barcelona's red light district that the City Council resolved to submit them to an 'artistic test', as a way of reducing their number. These bizarre street artists, mutely and inscrutably defying the specta- tor to detect a flicker of life in their expressionless faces or their still, poised bodies, are magical figures of suspended animation. Occasionally the clink of a tourist's coin may, especially thrillingly, cause a marble boy or a golden angel to break into clockwork motion, to come alive. Children stand transfixed by the wonder of a human, metamorphosed into a statue, transforming itself into a machine. Las Ramblas has long been a site for such entertainments. It was home for nearly four decades to the spectacular Roca Museum, established around 1900, whose col- lection included a mechanical wizard, displays of both beautiful and horrific waxworks, and a section devoted to human freaks, alongside embryological and anatomical models that could be opened up for examination, and were accurate enough to be used as both formal medi- cal and popular teaching tools. As the Wellcome Collection's recent exhibition in London of waxwork models amply attests,[1] earlier audi- ences were as excited and intrigued as modern-day tourists and museum visitors by the spectacle of models that are life-like and bodies that are machine-like.

Interest in the human–machine interface developed in the eight- eenth century with the invention of the first biomechanical automata, the most famous of which were Vaucanson's defecating duck and the flute-player. Anatomy and dissection played an important role in the building of these life-like automata, with anatomists producing flayed

1

specimens or écorchés for study, versions of artificial life fashioned from preserved human cadavers.[2] Later in the century, another version of artificial life attracted attention in the shape of embalmed cadavers whose life-likeness suggested the triumph of science and technology over death. In 1775 the anatomist John Sheldon displayed his God-like surgical and technical skills by transforming his young dead mistress into the most gazed-upon object in his extensive anatomical cabinet. A French visitor to London in 1784 has left us a description of Sheldon's 'entirely naked' and mummified mistress, whose glass-topped case occupied a 'distinguished place' in the anatomist's bedroom:

> She had fine brown hair, and lay extended as on a bed. The glass was lifted up, and Sheldon made me admire the flexibility of the arms, a kind of elasticity in the bosom, and even in the cheeks, and the perfect preservation of the other parts of the body. Even the skin partly retained its colour, though exposed to the air . . . Wishing afterwards to imitate the natural tint of the skin of the face, a coloured injection was introduced through the carotid artery.[3]

Palpating the embalmed body, Sheldon brings dead matter to life, troubling the cognitive boundaries normally separating life from death, the living from the merely lifelike, nature from art. In what seems another form of suspended animation, the dead mistress's flesh is firm and resilient; like living tissue it defies putrefaction. For Sheldon as Pygmalion, the body of this young woman represents the striving for eternal life through human artistry; like an eighteenth-century sex-doll, she also engenders an erotic fantasy of desire and animation.[4]

With a shortage of bodies for the growing number of medical students, embalmment was one way of meeting the demand for dissection. Although Sheldon's mistress was not preserved for this purpose, his wife later complained that some of their unsupervised visitors dealt 'roughly' with the corpse, cutting the threads 'to inspect the Inside'.[5] Here, with her inner mechanics exposed, the mistress's cadaver resembles an anatomical wax sculpture, or 'Venus' (see Figure 1). The year 1775 witnessed another bizarre case of embalming, performed by the celebrated anatomist William Hunter on the body of Mrs Van Butchel. From the day she died and for almost two months, Martin Van Butchell described in detail the cleansing and surgical processes of embalmment to which his wife was subjected. He was an enthusiastic assistant, and when the mummification was complete, he bore his wife's body home in triumph.[6] The case became something of a cause célèbre, with crowds

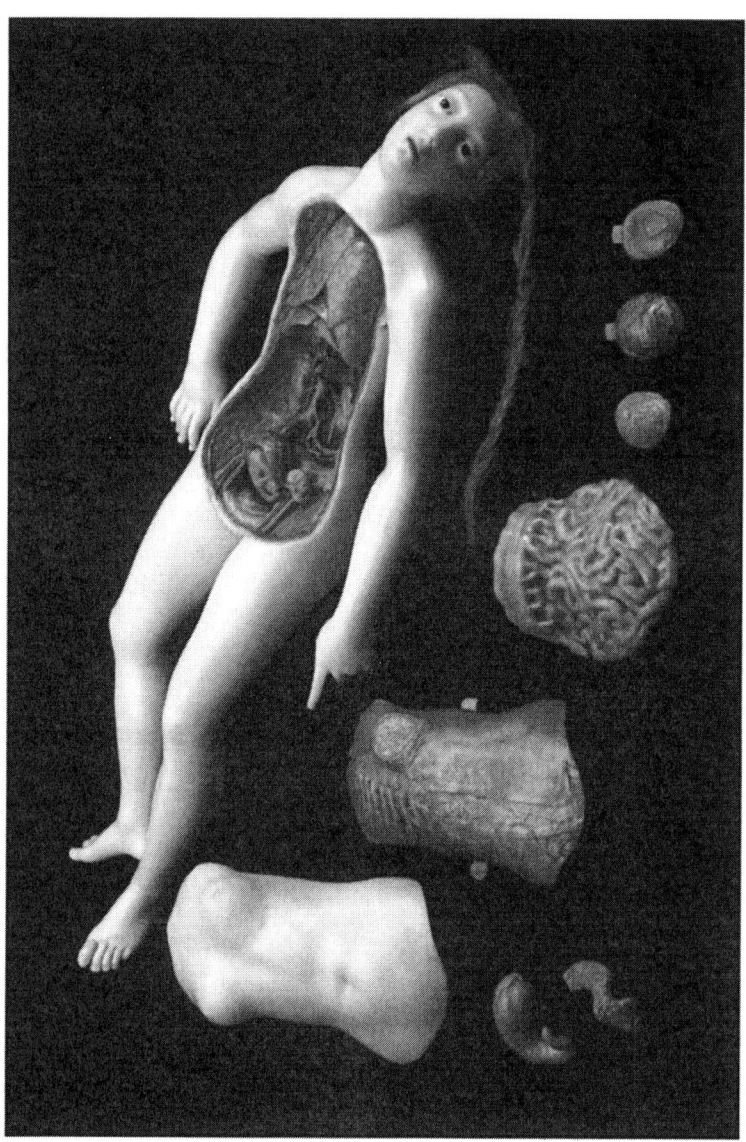

Figure 1: Wax 'Anatomical Venus'

of visitors hoping for a glimpse of Mrs Van Butchel's corpse, touted as sporting a much more spirited appearance in death than when alive.[7] One newspaper Epitaph praised Hunter for his supreme technical and artistic skills which had not just arrested time but reversed it, restoring Mrs Van Butchell to the charms she possessed before illness preyed upon them:

> . . . Hunter's skill in spite of Nature's laws,
> Her beauties rescued from Corruptions jaws;
> Bade the pale roses of her cheeks revive
> And her shrunk features seem again to live.

With his re-vivified wife at home, the devoted widower could now 'Caress, touch, talk to, even sleep/Close by her side'. More salaciously, the Epitaph suggests that Mrs Van Butchell is now as 'Firm, plump and juicy as before'; indeed, she is even 'fairer' and 'sweeter', and certainly more 'tractable'. With a touch of misogyny, Van Butchell is congratulated on laying claim to a woman 'whose temper's every day the same!'. At the end of life, art begins: the aesthetic process of embalmment makes death the ultimate beauty, with the 'true-to-life' corpse exceeding the living but dying Mrs Van Butchell. A taxidermic masterpiece, she is now the ultimate cabinet of curiosity.

Human statues which come to life like clockwork, and embalmed corpses more alive than the living: such liminal figures should not be relegated to the sideshows of history. Instead, they have a central place not only in the history of anatomy and medicine, but in our very conceptualisation of what it is to be human. This is why we feel such a frisson when we gaze upon or think about these categorically ambiguous forms, for their indeterminacy reminds us of the relations of propinquity between us and things. As Merleau-Ponty memorably formulates it,

> The visible world and the world of my motor projects are both total parts of the same Being ... Visible and mobile, my body is a thing among things; it is one of them. It is caught in the fabric of the world, and its cohesion is that of a thing.[8]

The human body, with its motor skills and its moving parts, shares a special kinship with the machine, and indeed there has long been a fascination with the unstable boundaries between them. The technological imagination has, since the Renaissance, widely permeated literary, visual, political and philosophical culture. Early modern writers and visual artists engaged with new technologies ranging from engines to microscopic lenses, from the camera obscura and calculating machines to automata. In the eighteenth century attempts to understand life by reproducing it mechanically resulted in life-like automata. But it is in the nineteenth century that the relationship between the human and the machine under post-industrial capitalism becomes a pervasive theme. From Blake on the mills of the mind by which we are enslaved, to Carlyle's and Arnold's denunciation of the machinery of modern life, from Dickens' sooty fictional locomotive Mr Pancks, who 'snorted and sniffed and puffed and blew, like a little labouring steam-engine', and 'shot out ... cinders of principles, as if it were done by mechanical revolvency',[9] to the alienated historical body of the late-nineteenth-century factory 'hand' under Taylorization, whose movements and gestures were timed, regulated and rationalized to maximize efficiency, we find a cultural preoccupation with the mechanization of the nineteenth-century human body that uncannily resonates with modern dreams and anxieties surrounding technologies of the human.

Cultural historians have begun to explore the discursive fields that have developed around key new technologies over the past three centuries and more, and how they have intersected with and shaped new forms of knowledge, self-representation and being in the world. Jonathan Sawday, for example, followed up his fine study of the early modern culture of dissection, *The Body Emblazoned* (1995), in which he argues that anatomy, by literally opening up the human body to scrutiny, produced the idea of human interiority, with a magisterial book on the imaginative history of the machine: *Engines of the Imagination: Renaissance Culture and the Rise of the Machine* (2007). Jonathan Crary, in his influential book *Techniques of the Observer* (1990), is concerned with 'vision and its historical construction' in the nineteenth century.[10] He argues that, alongside developments in physiological optics and optical technology, 'a new kind of observer took shape in Europe'[11] in the early decades of the nineteenth century, displacing the disembodied visual experience of the seventeenth and eighteenth centuries, defined by the incorporeal relations of the camera obscura, with an emphatically corporeal visual subject. Just as the camera obscura, he contends, was 'part of a larger organisation of representation, cognition and subjectivity in the seventeenth and eighteenth centuries',[12] so various new optical devices invented in the nineteenth century, such as the stereoscope, the kaleidoscope, the phenakistiscope and the diorama, were expressions of a new model of vision, whereby the observer comes to embody the optical effects of the machine, such that 'both were now contiguous instruments on the same plane of operation'.[13] New technologies of writing, information and communication – such as telegraphs, telephones and typewriters – have also yielded rich seams of speculation about the relays between the natural and the technological. For cultural theorist and literary critic Mark Seltzer, looking at the shift from handwriting to typewriting in late-nineteenth century American machine culture, what is notable is not the replacement of bodies by machines but the radical and intimate coupling of bodies and machines.[14] Nietzsche expressed this coupling succinctly. The first philosopher to use a typewriter, he typed in one of his letters: 'Our writing tools are also working on our thoughts'.[15] Finally, for Mary Ann Doane in her book *The Emergence of Cinematic Time* (2002), it is the new technology of film that is of interest. Alongside the pocket watch, which Georg Simmel in 1903 linked to a new preoccupation with temporal precision that was crucial to 'the technique of metropolitan life', and the standardization of time demanded by rail travel and telegraphy, Doane argues that the cinema was a major factor in the reconceptualization of time in modernity.[16]

The essays collected here continue this critical trajectory, offering detailed analyses of particularly compelling constellations of minds, bodies and machines in the nineteenth-century cultural imaginary. The first two chapters explore, respectively, ideas about 'imagination-machines' and 'influencing-machines' that emerged at the beginning of the century, the moment when the machine began to invade – and for some seems to reflect, define or even determine – human consciousness. In Chapter 1 of this book, Peter Otto's focus is on Gothic fiction, and his interest, like Crary's, is in the cultural meanings of the camera obscura in the late eighteenth and early nineteenth centuries. He takes his start from Coleridge's contempt for the Gothic genre whose fantastic effects are 'supplied *ab extra* by a sort of mental *camera obscura* manufactured at the printing office', leading to a confusion between *ab extra* projections of reality and *ab intra* delusions of imagination in the mind of the reader. Through his readings of a Gothic novel that exemplifies Coleridge's despised imagination-machines, Ann Radcliffe's *The Mysteries of Udulpho* (1794), Otto traces how the real-unreal worlds and affective power of the genre were shaped by the paradigm of camera obscura, and were moreover an inevitable product of the machinic consumer culture that was to define the long nineteenth century, not least its publishing industry.

John Perceval, one of the figures Steven Connor discusses in Chapter 2 on early nineteenth-century elaborations of the 'influencing machines' postulated by sufferers from schizophrenia, similarly invoked the printing office when he drew an analogy between a printing press and an imagined mechanical breathing regulator that enables an orderly production of thoughts. The mechanization of print media was an underlying trope for many of the nineteenth-century writers discussed in this volume, including Connor's exponents of a machinery of the mind. In his article 'On the Origin of the "Influencing Machine" in Schizophrenia', the pioneer psychoanalyst and neurologist Viktor Tausch (1879–1919) was the first to name the controlling engines imagined by some paranoid schizophrenics as responsible for their psychoses. Connor begins his own exploration of the phenomenon with the first documented case, that of James Tilly Matthews, who obligingly made a drawing of the machine – the Air-Loom – to which his mind was subject, complete with a key to all its components (see Figure 3 in Chapter 2). This machine becomes an allegory of the process of thought, rendering the operations of mind in strangely material terms. The pneumatic components of these ingenious engines of thought are of particular interest, their fluid mechanics a delicious parody of the ethereal vital fluids associated with mesmerism, animal magnetism and electricity. Connor thus joins Iain McCalman (Chapter 6) in exploring the curious overlap between scientific and esoteric thought, incorporating a discussion of the way in which air machines assist in the corporealisation of thought. Connor concludes by tracing the delusional pneumatic visions of other nineteenth-century sufferers from this psychiatric disorder, arguing that '[t]he passion of the machine arises not in the fear of alienation by, but in the voluptuous delight of identification with it'.

Connor observes that Perceval, in analysing his own paranoid condition, defines madness as 'the incapacity to distinguish literal from figurative expressions, and the consequent tendency to take metaphors literally'. As our next cluster of essays, by Katherine Inglis, Paul Crosthwaite and Marie Banfield, makes clear, though, the machine madness Perceval identifies among the psychologically ill could be said to be symptomatic of a broader cultural phenomenon in which the categories of human and machine were unsettled, and in which there was a propensity to take mechanical metaphors literally. In Chapter 3, on James Hogg's *The Three Perils of Woman* (1823), Inglis focuses on the figure of the 'ghastly automaton' of the mother, and the conjunction of maternity, madness and mechanization in a novel in which the maternal body is represented as a defective machine. She notes that the meanings of 'automaton' are unstable and contradictory, both within and beyond the text: an automaton is 'at once a true self-mover and a mimic of autonomous motion, a mechanical simile that moves *as if* endowed with life'. And yet she observes of Hogg's heroine Gatty, who gives birth in an asylum during a three-year period of unconscious-ness, that the metaphorical association of woman with machine goes beyond that of simile; as Inglis argues, Gatty 'becomes an automaton indeed, not *like* an automaton'. We are reminded equally of Connor's machine-mad literalists and of Otto's Gothic 'imagination-machines', with their confusion of reality and delusion. Furthermore, Inglis locates Hogg's Gothic mechanization of childbirth within eighteenth and early nineteenth-century medical culture, recalling the anatomical models gathered for the recent Wellcome exhibition, so many of which were pregnant females, with a special focus on the foetus *in utero* and the potential complications of childbirth. In the context of the emergence of obstetrics and the suppression of female-dominated midwifery, Inglis argues that the representation of the maternal body as a machine in eighteenth-century midwifery literature and medical illustration, and the use of androids/automata in medical instruction, reduced childbirth to mechanical principles and the mother to an engine of reproduction.

Like Inglis, Paul Crosthwaite (Chapter 4) is intrigued by the slip-pages between metaphor and simile in the relationship between human and machine, but his own interest is in the cultural history of computer-generated writing, and in particular late-eighteenth and early nineteenth-century mechanical writing figures. He modifies Jessica Riskin's distinction between an eighteenth-century ethos of simulation and a nineteenth-century culture of analogy by arguing that through-out both periods the makers of writing automata gave their machines a human form and human movements in order to create the illusion that they were autonomously rather than mechanically generating the words they produced – that they were not just *like* authors, they *were* authors. The very word 'author', from the Latin 'auctor', creator, denotes an originary function, and therefore, as in the case of mechanical child-birth, the concept of a clockwork author seems especially paradoxical, the more so when it coincides with the emergence of a Romantic dis-course of authorship. Of course, as Crosthwaite notes, '[t]he author of Romantic theory is a notoriously contradictory being'; and the ambigu-ous agency of these writing automata could be said to resonate with Wordsworth's celebration in 'Tintern Abbey' of 'all the mighty world/ Of eye and ear, both what they half-create,/ And what perceive' – and also, we might add, with Coleridge's Aeolian harp, as well as with the relational confusion of *ab intra* and *ab extra* projection he identifies as the product of 'imagination-machines' in Gothic writing. Such paral-lels notwithstanding, in the age of digital as in the age of mechanical reproduction there is, Crosthwaite argues, something ineluctably human about the writing and the reading of poetry that eludes the automaton-author and equally the cybernetic poet: an intention to *mean* that 'can arise only from embodied existence in the referential realm of material objects and relations'.

This takes us back to metaphor, and forward to Marie Banfield's discussion (Chapter 5) of 'Metaphors and Analogies of Mind and Body in Nineteenth-Century Science and Fiction' which, as if to confirm Riskin's thesis, begins with William James's remark that science is 'nothing but the finding of an analogy'. Banfield's focus is on the work of the psychologists Alexander Bain, George Henry Lewes, James Sully and James himself, and on the novelists who most engaged with contemporary psychology, George Eliot, Henry James and George Meredith. Nineteenth-century writers, she argues, were drawn to metaphor because of its capacity to juxtapose different discourses and domains, but also because it enables ambivalence in the face of human paradox and contradiction. However, Lewes warned of the dangers of 'taking a metaphor for a fact, and arguing as if the mind were a *mirror*'. That way, as we know from Connor's essay, madness lies. The mind does not passively offer a 'faithful reflection of the world', only 'a faithful report of its own states'. Echoing Wordsworth, Lewes sees the mind as 'the partial creator of its own forms'. For William James, similarly, '[t]he knower' is not a mere reflective mirror, but is 'an actor, and co-efficient of the truth on one side, whilst on the other he registers the truth which he helps to create'.

The mind and the emotions are, in Bain's psychology, profoundly related to the physiological body, and both mind and body are described in terms of metaphors drawn from contemporary science and technology. His model of consciousness, for example, is informed by energy physics and thermodynamics. Banfield demonstrates how the representation of character and consciousness in the fiction of Eliot, Meredith and Henry James is inflected by such metaphors in ways that suggest new understandings of the interdependence of the psychological and the corporeal, and a new conceptualization of consciousness as a dynamic force or current that is part of the objective world of things. The eponymous protagonist of Meredith's *Diana of the Crossways* (1885), for instance, herself a writer, has a mind like 'a steam-wheel', a metaphor which, as Banfield notes, 'signifies both a process of relentless, mechanical activity and the work and productivity of the brain'. Like an automaton, she is at once genuinely self-powered and mechanically driven. Furthermore, '[m]etaphors', the reader is told, 'were her refuge'. 'The banished of Eden had to put on metaphors ... Especially are they needed by the pedestalled woman in her conflict with the natural'.

If pedestalled women needed metaphors in their conflict with the natural, then so did those excluded by their class from the cultural and scientific mainstream, such as the self-educated plebeian radical Alfred Wallace, the co-discoverer with Darwin of natural selection, and the subject of Iain McCalman's essay (Chapter 6). Our next section is devoted to some notable clashes between mid-nineteenth-century scientists that throw into focus some of the key tensions in their thinking about the relations between minds and bodies, and between human and inorganic behaviours. Continuing our theme of emergent theories about the mind and consciousness in the new field of psychology, McCalman explores the vexed question of Wallace's conversion to spiritualism, and his theory of the spiritual evolution of mind. In a decisive break with Darwin, some ten years after the publication of *On the Origin of Species* (1859), Wallace published a devastating paper entitled 'On the limits of natural selection as applied to man' (1870), in which he postulates a higher 'Intelligence' in order to account for man's mental descent from animals and the growth of human consciousness. Seeking out the 'common social context' that makes it possible to reconcile Wallace's adoption of a spiritualist interpretation of mental evolution with his belief not only in natural selection but also in the secular ideologies of Owenism and phrenology, McCalman finds it in his socio-political background. As a plebeian autodidact and radical, Wallace was introduced to phreno-magnetism, a new pseudo-scientific phenomenon which 'offered him a science of human behaviour that linked the bodies, minds and environments of man in law-like relationships, yet also allowed room for elements of irrationalism, or at least of sentience and emotion'. Understanding this context, McCalman argues, enables us to see Wallace's spiritualist turn not as the volte face of a confirmed materialist but, aptly enough, as an adaptation of his earlier thought.

Wallace has traditionally been represented as a divided man: at once the explorer-naturalist who was 'a man of science' and a social philosopher who was 'a man of nonsense'. If McCalman overturns the view that he was a man of nonsense, Daniel Brown (Chapter 7) contests the very idea that science and nonsense are polarized categories by making a case for the principle of Nonsense, as it was espoused by James Clerk Maxwell in his poetic ripostes to fellow physicist John Tyndall's reductionism in his mechanistic construal of atomism:

> Yield, then, ye rules of rigid reason!
>   Dissolve, thou too, too solid sense!
> Melt into nonsense for a season,

      Then in some nobler form condense.
    Soon, all too soon, the chilly morning,
      This flow of soul will crystallize,
    Then those who Nonsense now are scorning,
      May learn, too late, where wisdom lies.

Tyndall was one of the leaders of a group of London-based scientists known as the 'Metropolitans', who in the 1860s and 1870s espoused a radical form of scientific materialism that was inaugurated by his 1868 Norwich lecture to the British Association for the Advancement of Science. In this lecture, he argues that atomic forces constitute a natural mechanism that overturns the argument from design by demonstrating nature's independence of a creator: for, like William Paley's watch, natural phenomena 'also have their inner mechanism, and their store of force to set that mechanism going'. Maxwell, a member of a rival group of 'North British' physicists, took issue in his own addresses to the British Association with what he perceived to be Tyndall's reductive materialism, but he also addressed their differences in a more playful way in a series of ribald verses he penned in the early 1870s for the Red Lion Club. By contrast with Tyndall's mechanistic determinism, Maxwell and his colleagues developed a statistical theory which, Brown explains, 'recognises that molecules have different properties from the bodies they compose and that their mechanistic behaviours can interact in unforeseen ways'. Brown's focus in this essay is on these comic poems in which, he argues, Maxwell parodies Tyndall's scientific materialism, invoking Lucretius' poem on atomism, *On the Nature of Things*, in particular the doctrine of free will that is so integral to his physics, in order to propose an alternative cosmology.

Maxwell's verses from the early 1870s play upon the anthropomorphic character of his physical theory; indeed, as Brown argues, they are premised upon the implication of a radical analogy between human and inorganic bodies and behaviours. Moving from molecules to microbes, in our next essay, on 'Influenza, Informatics, and the Body in the Late Nineteenth Century', James Mussell (Chapter 8) considers the extension of biological language to electronic communication systems. Beginning with a contemporary analogy between computer viruses and pandemics, and reflecting upon the spread of both information and infection, Mussell takes us back to the influenza pandemics of the late nineteenth century which, he argues, were simultaneously biological and cultural phenomena, 'able to infect both the human and the social body simultaneously and move between both'. Like Maxwell's molecules, the microbe was the subject of both comic lyrics and anthropomorphic fantasy. Florence Fenwick Miller, for example, writing in the *Illustrated London News*, reflects upon the 'somewhat appalling idea that each human system forms a world, in which a whole myriad of microscopic animalculae are born, live by their own exertion, perhaps form kingdoms or republics, hoard wealth, prey on each other, rear offspring, and depart from life', and imagines how 'the enterprising microbe has travelled with speed, perhaps by train or by telegraph, to other parts of the kingdom'.

Focusing on how viruses move through both human and social bodies, Mussell captures in the 'awful poetry' of microbiological migration a host of anxieties swirling around the porousness of things, the permeability of bodies, and the materiality of language. In so far as these anxieties are fundamentally identified with modernity, Mussell's essay joins with the final essay in our collection, Laura Salisbury's 'Linguistic Trepanation: Brain Damage, Penetrative Seeing, and a Revolution of the Word' (Chapter 9), to investigate the way in which the latest technologies offered a new ontology of the human mind, body and perceptual system. For some early twentieth-century writers, bodily penetration, whether by a soldier's bullet, a surgeon's knife, or the radiant waves of an x-ray, come to signify modernity. For Guillaume Apollinaire, for example, Salisbury argues that his head wound and surgically trepanned skull become 'a site of interpenetration carved out by the force of war that brings minds, bodies and machines into a commerce with one another that initiates linguistic innovation'. This exciting and unsettling sense of the porosity of the body, and of the unstable boundaries between cognition, language, the physicality of the human brain and technology, recalls an earlier historical moment when the French physician and neuroanatomist Paul Broca, and after him the German anatomist and neuropathologist Carl Vernicke, as part of their research into aphasia, developed materialist models of language production based on morphological studies of the structure of the brain and the localization of its functions. Salisbury traces how their work intersected with what she describes as a 'general discursive environment' in the latter half of the nineteenth century, in which bodies were imagined as fleshly machines, and how '[t]he linguistically trepanned revolutionary of the word paradoxically uses a technological imaginary to speak and write beyond the structures and shapes of the nineteenth-century machine'.

For these later writers, working around the turn of the twentieth century, like the figures with whom this volume begins at the turn of the nineteenth century, corporeality and technology are conceived of as existing in profound metaphorical relationship. But the nature of the human–machine interface has changed. Instead of automata and steam engines, the analogies for a technology of the human invoke the electrical impulses of the telegraph, the phonograph, cinema and the x-ray rather than the galvanized corpse; the microbe and the molecule rather than the insect; the nervous system rather than the brain-machine. Today, another hundred years on, the metaphorical distance between human and machine could be said to have collapsed altogether, as can be seen, for instance, in the case of Kevin Warwick, Professor of Cybernetics at the University of Reading and controversial author of *I, Cyborg* (2002). Warwick believes that, given our creation of highly intelligent machines, we must keep pace by assimilating elements of this technology into ourselves so that we become cybernetic organisms, part-human, part-machine. An important part of the cyborg project involves, for Warwick, 'the realization of the superhuman being'.[17] With an implant array fired onto his median nerve, Warwick thrills to the sensation of transmitting brain signals via a computer onto a robot hand. Achieving nervous system to nervous system communication forms an even more ambitious stage in the upgrading of the human. With his wife Irena, Warwick investigates a 'pure' form of communication via a 'his-and-hers implant experiment', a connection which goes beyond movement to the 'emotional signals associated with being happy, depressed, shocked or even sexually aroused'.[18] In so far as Warwick's project involves 'jacking into' his wife's nervous system by means of direct neural links, his cyborg project gives a new spin to the ironic and revolutionary cyborg of Donna Haraway's feminist-socialist 'Manifesto for Cyborgs' (1985), an essay which opened a space for new forms of identity – 'we are all chimeras, theorized and fabricated hybrids of machines and organism'.[19] Although Warwick's cyborg project has none of the irony of Haraway's, both theorist and practitioner share a daring flirtation with 'transgressed boundaries, potent fusions, and dangerous possibilities'.[20]

With her 12 pairs of legs, Aimee Mullins, fashion model and Paralympian, puts all the irony back into Haraway's cyborg.[21] A double amputee since the age of one, Mullins uses her celebrity and great physical beauty to be the figurehead of the prosthetic industry, drawing in animatronic designers, aerospace engineers, glass blowers, wood-workers and all kinds of other artisans to design and build legs which fuse art with cutting-edge technology, playfully confounding the boundaries of animal and human, human and machine (see Figure 2). Turning the loss of both lower limbs into a triumph, Mullins offers the prosthetist

Figure 2: Aimee Mullins, in medical corset with sutures, flamenco skirt and hand-carved wooden legs, solid ash with grape-vine and magnolia motifs, by Alexander McQueen

carte blanche to design matching female 'fantasy' legs, many of which, as wearable sculpture, fulfil some whimsy or feminine ideal. There are Mullins' 'pretty legs', for instance, impossibly long, thin and beautiful according to the Barbie doll ideal. These make her 'fashion-able' while her famous carbon-fibre 'sprinting legs' and 'cheetah' feet, with which she has broken world records, transform her into the super-able athlete, the embodiment of superwoman prowess. Keen to challenge prevailing attitudes concerning 'the human', Mullins pits the artificial against the natural, invoking poetry and the imagination to suggest that the process of transformation through artifice is illimitable. In doing so her arguments resemble those of Sir Philip Sidney who praised the poet as the only one who 'disdains to be subjected to nature'; lifted up by the vigour of his own invention, and 'freely ranging only within the zodiac of his own wit', the poet makes things either better than nature, or makes them 'quite anew, forms such as never were in Nature'.[22]

In E. T. A. Hoffman's tale of 'The Sandman' (1816), the association of femininity with artificiality is an uncanny and disturbing one, leaving many men anxious that their lovers are automata. To convince themselves otherwise these men 'would request their mistress to sing and dance a little out of time, to embroider and knit, and play with their lap-dogs, while listening to reading, &c'.[23] The negative suggestion that woman is an inherently unnatural form, implied, for instance, by Hogg's machinic maternal body, gets exploded by Mullins who emphasizes the power of technology to establish a new measure of the human, both male and female. For Mullins, prosthetic legs are not about replacing loss or overcoming deficiencies; like Warwick, she is focused on cyborg-like augmentation and upgrading, and on redefining herself through new forms of post-human embodiment and identity. In the prosthetics industry, where technology fuses with fantasy and robotics with poetry, Mullins uses her beauty to create legs out of 'glass' (transparent polyethylene), solid ash, brass and dirt – 'legs' which have never been seen before.

The unanswerable, ever-vexatious question as to what is the measure of the human, now that prostheses have become so sophisticated, can be seen in the sporting world's disputes surrounding Oscar Pistorius, the South African Paralympic runner known as the 'fastest man on no legs'. A double amputee world record holder, Pistorius has been competing against able-bodied athletes since 2007 and is now determined to participate in the next summer Olympics in London in 2012. His 'Cheetah' legs are controversial, however, generating claims that they exceed the capabilities of biological legs, thus giving him an unfair advantage over able-bodied runners. On these grounds the International Association of Athletics Federations ruled him ineligible to compete in the Beijing Olympics in 2008, only to find its decision overturned a few months later by the Court of Arbitration for Sport.

The progressive, utopian cyborgization of Mullins and Pistorius, evident in the substitution of flamboyant and controversial prostheses for their missing body parts, is a long way from the eighteenth-century project of building automata which simulated life as closely as possible.

Today the prosthetic industry is focused on the new science of understanding human adaptability in the face of disability or loss. In a recent workshop staged by the MIT Media Lab, entitled 'h2.0: new minds, new bodies, new identities',[24] Oliver Sacks spoke movingly of the human brain's hunger to map itself onto the world; the phantom limb's longing for re-embodiment was, he claimed, like a 'soul longing for its body'. The metaphor of the soul helps us to understand the mysterious mechanism of this cerebral hunger with a view to marshalling its forces. Understanding the retention of sensory and motor powers despite the loss of a limb, and the fluent intermodal transfer of senses when one sense is impaired (the recruitment of the blind person's visual cortex by hearing, for instance) promises to usher in an era of smart prostheses at the neural–digital interface – technologies which will merge with our bodies and our minds to change our concept of human capability forever.

Computational theories of mind are now behind us when it comes to the building of intelligent machines. The mind, according to David Bates, 'is less a processor of information than a creator and interpreter of associative relationships. A focus on symbolic logic has given way to an interest in how mental activities such as metaphor, narrative, and analogy can help reveal the essence of human reason'.[25] Similarly, a formulation of human intelligence which erases embodiment, emotion and sociality, typified for many by the Turing test of 1950, no longer seems plausible.[26] This trend to embodiment can be seen in the pursuit of building sociable robots capable of monitoring the health of children or the elderly; or the invention of everyday artefacts that can read our intentions so that they can collaborate with us. Turing's optimism that one day a disembodied machine would be invented which would 'compete on equal terms' with 'any one of the fields normally covered by the human intellect', including the writing of a sonnet, is now no longer credited.

It is during the nineteenth century, the age of machinery, that we begin to witness a sustained exploration of issues to do with minds, bodies, machines. This book, which opens with the imagination-machines of late eighteenth-century Gothic fiction and closes in the trenches of the First World War, brings together a collection of essays whose overall orientation is literary and historical but which also touches on philosophy, the history of medicine and psychiatry, new technologies and their impact upon conceptions of human cognition, man/machine boundaries, the boundaries between materialist and esoteric sciences, and the transformation of our ideas of the natural human form. Laura Salisbury builds her chapter around Apollinaire's suggestive idea that 'Wise men ceaselessly investigate new universes which are discovered at every crossroads of matter'. A century later, in the context of a new digital and cybernetic revolution, we find ourselves reflecting on new 'crossroads of matter', and their implications for minds, bodies, machines.

Notes

1. 'Exquisite Bodies', 30 July–18 October 2009, Wellcome Collection, London; see www.wellcomecollection.org/whats-on/exhibitions/exquisite-bodies.aspx
2. For the link between anatomy and automaton-making, see Joan B. Landes, 'The Anatomy of Artificial Life: An Eighteenth-Century Perspective', in Jessica Riskin (ed.), *Genesis Redux: Essays in the History and Philosophy of Artificial Life* (Chicago: Chicago University Press, 2007), 96–116.
3. The Frenchman was B. Faujas de Saint Fond. His account is to be found in *A Journey through England and Scotland to the Hebrides in 1784*, 2 vols (Glasgow: Hugh Hopkins, 1907), 41–46. See also Jessie Dobson, 'Some Eighteenth-Century Experiments in Embalming', *Journal of the History of Medicine* 8 (October, 1953), 437.
4. See Patricia Pulham, 'The Eroticism of Artificial Flesh in Villiers de L'Isle Adam's *L'Eve Future*', in Deirdre Coleman and Hilary Fraser (eds), 'Minds, Bodies, Machines', *19: Interdisciplinary Studies in the Long Nineteenth Century*, 7 (2008).
5. Dobson, 'Some Eighteenth-Century Experiments', 438.
6. Thomas Joseph Pettigrew, *A History of Egyptian Mummies* (London: Longmans, 1834), 258–9.
7. The stream of spectators became so pressing that Van Butchell put up a notice stipulating that strangers could enter only if introduced by a friend; see Dobson, 'Some Eighteenth-Century Experiments', 436.
8. Galen A. Johnson (ed. and trans.), Michael B. Smith (translation editor), *The Merleau-Ponty Aesthetics Reader: Philosophy and Painting* (Evanston: Northwestern University Press, 1993), 123–4.
9. Charles Dickens, *Little Dorrit* (1847; London: Penguin, 1967), 190, 202.
10. Jonathan Crary, *Techniques of the Observer: On Vision and Modernity in the Nineteenth Century* (Cambridge, MA.: MIT Press, 1990), 1.
11. Crary, *Techniques of the Observer*, 6.
12. Jonathan Crary, 'Techniques of the Observer', *October*, 45 (1998), 3.
13. Crary, *Techniques of the Observer*, 129.
14. Mark Seltzer, *Bodies and Machines* (New York and London: Routledge, 1992), 12–13.
15. Quoted by Friedrich A. Kittler, *Gramophone, Film, Typewriter* (Stanford: Stanford University Press, 1999), 200.
16. Mary Ann Doane, *The Emergence of Cinematic Time* (Cambridge, MA: Harvard University Press, 2002).
17. Kevin Warwick, *I, Cyborg* (Urbana and Chicago: University of Illinois Press, 2002), 178.
18. Warwick, *I, Cyborg*, 138.

19. Donna Haraway, 'A Cyborg Manifesto: Science, Technology, and Socialist-Feminism in the Late Twentieth Century', in *Simians, Cyborgs and Women: The Reinvention of Nature* (New York: Routledge, 1991), 150.

20. Haraway, 'A Cyborg Manifesto', 154.

21. For Mullins' most recent appearance on TED, see www.ted.com/talks/aimee_mullins_prosthetic_aesthetics.html

22. Sir Philip Sidney, *An Apology for Poetry: or, The Defence of Poesy*, ed. Geoffrey Shepherd (Manchester: Manchester University Press, 2002), 100.

23. E.T.A. Hoffmann, 'The Sandman', in John Oxenford and C. A. Feiling (trans.), *Tales from the German, Comprising Specimens from the Most Celebrated Authors* (London: Chapman and Hall, 1844), 163.

24. 'MIT Media Lab: A One-Day Symposium', May 9, 2007, at MIT's Kresge Auditorium (Hosts: Hugh Herr and John Hockenberry). See http://h20.media.mit.edu/.

25. David Bates, 'Creating Insight: Gestalt Theory and the Early Computer', in Jessica Riskin (ed.), *Genesis Redux: Essays in the History and Philosophy of Artificial Life* (Chicago: Chicago University Press, 2007), 237–59.

26. For a sympathetic reading of Turing's work which situates it within the specific historical and affective conditions of early cybernetic research, see Elizabeth A. Wilson, 'Imaginable Computers: Affects and Intelligence in Alan Turing', in *Prefiguring Cyberculture: An Intellectual History* (Cambridge, MA.: Power Publications and MIT Press, 2002), 38–51.

1
Inside the Imagination-Machines of Gothic Fiction: Estrangement, Transport, Affect

Peter Otto

> Lo, here a CAMERA OBSCURA is presented to thy view, in which are lights and shades dancing on a whited canvas, and magnified into apparent life! . . . walk in, and view the wonders of my INCHANTED GARDEN.
>
> Erasmus Darwin,
> *The Loves of the Plants* (1789)

Coleridge remarked in disgust that the entire '*materiel* and imagery' of gothic fiction's waking dreams are 'supplied *ab extra* by a sort of mental *camera obscura* manufactured at the printing office, which *pro tempore* fixes, reflects and transmits the moving phantasms of one man's delirium, so as to people the barrenness of an hundred other brains'.[1] The *ab extra* projections of the camera obscura are normally contrasted with the *ab intra* projections of the magic lantern. The first provides a figure for realistic representation; the second for the delusions of imagination. And these poles normally provide the points of reference in relation to which gothic narratives are mapped and interpreted by their modern readers. Yet according to Coleridge, in gothic fictions the opposition between *ab extra* and *ab intra* projections is confused by an array of paper imagination-machines able to project, into the mind of the reader, *ab intra* delusions as if they were *ab extra* projections of reality. This imagination-machine not only multiplies phantasms, annexes the mind to itself (as the screen/auditorium within which its phantoms come to life) but also, through the passions they arouse in readers, gives these phantasms purchase on real bodies and the real world.

Coleridge is of course attempting to dispatch the genre with its own poison, by telling a gothic story about gothic fictions. However for

19

readers of gothic fictions, the poison he manufactures is likely to have provided the cure already for a mundane existence, and as such to have been what drew them to these fictions in the first place. In gothic fictions the marvellous, presented in the guise of the actual, gained the degree of reality necessary for it to escape the regulative machinery of inner sense, establish an improbable reality that seemed real, and so provoke 'a constant vicissitude of . . . passions' that readers found 'interesting'[2] precisely because they seemed not to have been aroused by (the reader's) inner or outer reality. Far from being ashamed of the mechanisms that produce apparently-real presence, as Coleridge implies they ought, gothic fiction parades them, delighting in the revelation that what had been experienced as real presence is produced by textual, cultural or perceptual machines.

This chapter takes Ann Radcliffe's *The Mysteries of Udolpho* (1794) as an exemplary instance of the imagination-machines described by Coleridge. After a discussion of the camera obscura and its roles in the seventeenth and eighteenth centuries, it focuses on three episodes from *Udolpho* that explore the experience of reading gothic fictions: the journey of Emily St Aubert (the heroine of Radcliffe's novel, often described as a figure for the reader) from the gates of the castle of Udolpho to 'the double chamber', her bedroom in Udolpho; the night-time appearance in 'the double chamber' of Count Morano as tender lover, and the dangerous shadow-show that ensues; and Ludovico's mysterious transport from a locked room in the Chateau-le-Blanc (a recension of the fate avoided by Emily).[3]

In place of sociological or psychological readings of the (virtual or ideal) worlds conjured by these texts, I argue that Radcliffe uses the gothic to explore the implications for identity and epistemology of the recognition that we can be possessed by passions and thoughts (sparked by *ab intra* delusions that are experienced as if they were *ab extra* projections of reality) that are in an important sense 'new'. This shift of focus allows me to redescribe some of the key features of the genre: its unprecedented mixing of conventions designed to represent the actual world with those normally deployed to evoke the marvellous; its ability to evoke in readers a powerful sense of the reality of its unreal worlds; and the consequent power of these unreal-realities to rouse the emotions of those who enter them.

The camera obscura

In its simplest form, a camera obscura comprises a darkened chamber with an aperture (usually filled with a convex lens) in one of its walls.

This simple mechanism is sufficient to project onto the wall facing the aperture or onto an appropriately placed screen a two-dimensional pattern of light and shade, which the mind construes as a remarkably accurate (although inverted), apparently three-dimensional picture of the external world.

Leonardo da Vinci (1452–1519) and Johannes Kepler (1571–1630) were the first to develop a detailed analogy between the camera obscura and the human eye. It had previously been assumed, following the Greek physician Claudios Galenos (129–99), that the ocular lens was the screen on which pictures of the external world were formed and as such was the primary organ of vision.[4] In *Ad Vitellionem Paralipomena quibus astronomiae pars optica traditur* (1604), Kepler argued that 'the pupil acted like the pinhole of a camera obscura and that the admitted light was focussed by the crystalline humor to form an inverted image on the retina'.[5] From this time through the seventeenth and eighteenth centuries, the camera obscura was routinely used to describe the mechanics of vision. Isaac Newton (1642–1727), for example, notes towards the beginning of his *Opticks* (1704) that 'the Pictures of Objects [are] lively painted' on the back of our eyes. 'And these Pictures, propagated by Motion along the Fibres of the Optick Nerves into the Brain, are the causes of Vision'.[6] Understood in this way, the eye is an optical apparatus, a machine able to operate without the conscious intervention of the mind and to produce images even after the death of the person to whom it belongs.

The most famous and for the eighteenth century most influential use of the analogy between camera obscura and eye is found in John Locke's *An Essay on Human Understanding* (1690), where it provides an analogy for the operation of the senses in general, and for the relation between the data they produce and the mind. According to Locke,

> the understanding is not much unlike a closet wholly shut from light, with only some little openings left, to let in external visible resemblances, or ideas of things without: would the pictures coming into such a dark room but stay there, and lie so orderly as to be found upon occasion, it would very much resemble the understanding of a man, in reference to all objects of sight, and the ideas of them.[7]

The camera obscura is used by Locke to dismiss early-modern accounts of the mind and experience, and to compose a dramatically new picture of the relation between these terms. First, the camera obscura's screen provides a powerful figure for a mind unencumbered by innate ideas: we come into this world, Locke writes, with minds like 'white paper, void of all characters, without any ideas'.[8] Second, in contrast to early-modern accounts of the mind, where 'the fact of the brain's being of the body virtually eliminates the possibility of anything like a rational perception of the world', the camera obscura's inert walls conjure a mind housed within a void, set apart from the body *and* its passions.[9] And third, the machinic vehicle of perception, which delivers again and again the same finite set of simple ideas to the mind, provides a stable foundation for thought, which rises from (yet is unable to move beyond) these foundations. When Locke then adds to the camera obscura of perception a device able to retain the simple ideas projected onto the 'white paper' of the mind, and a Faculty (the Understanding) able to judge, compare and then combine them, he conjures a powerful impression of the autonomous, self-governing subject of modernity, whose identity is formed entirely through experience.[10]

In Jonathan Crary's account of this development, the camera obscura 'performs an operation of individuation', defining the 'observer as isolated, enclosed, and autonomous within its dark confines'; provides 'a figure for both the observer who is nominally a free sovereign individual and a privatized subject confined in a quasi-domestic space, cut off from a public-exterior world'; and displaces 'the observer's physical and sensory experience' with 'the relations between a mechanical apparatus and a pre-given world of objective truth'.[11] It is only 'in the 1820s and 1830s', according to Crary, that the observer begins to be repositioned 'outside of the fixed relations of interior/exterior presupposed by the camera obscura and into an undemarcated terrain on which the distinction between internal sensation and external sign is irrevocably blurred'.[12] According to this new model, vision is the product of 'arrangements of forces out of which the capacities of an observer are possible'; the observer is able 'to generate visual experience'; and vision consequently need 'not refer or correspond to anything external to the observing subject'.[13] All this seems straightforward enough; yet when one looks more closely at Locke's *Essay*, the camera obscura operates in ways that perplex Crary's sharp division between eighteenth-century and nineteenth-century modes of vision.

According to Locke, our senses 'convey into the mind several distinct perceptions of things', but this occurs only 'according to those various ways wherein those objects do affect them'.[14] The 'immediate object of perception, thought, or understanding' is 'sensations or perceptions in our understandings', rather than objects in the external world. We can therefore be certain of the simple ideas we receive but not of the reality they appear to represent.[15] It therefore follows, as David Hume (1711–76) remarks, that the 'ultimate cause' of 'those *impressions*, which arise from the *senses* . . . [is] perfectly inexplicable by human reason, and 'twill always be impossible to decide with certainty, whether they arise immediately from the object, or are produc'd by the creative power of the mind'.[16] Or as Thomas Reid (1710–96) concludes, 'no man can . . . show, by any good argument, that all our sensations might not have been as they are, though no body, nor quality of body, had ever existed'.[17]

In a camera obscura one can of course always step out from its 'dark room' to check whether its virtual landscapes correspond with those outside. This possibility conditions the association of the camera obscura with realism. Yet we can't step outside the camera obscura of perception. Consequently, if our senses were to change, our view of the world would also change. Locke therefore concedes the possibility that 'in other mansions' of the creation there may be other 'and different intelligent beings, of whose faculties [we have] as little knowledge or apprehension, as a worm shut up in one drawer of a cabinet hath of the senses or understanding of a man'.[18] In this admission one can glimpse a nascent gothic and romantic sensibility which dissolves the unified world of traditional metaphysics into multiple realities: the 'dark room' that defines the camera obscura of the worm exists within the larger space of the cabinet (another term used by Locke to describe the 'dark room' of the mind),[19] which is placed, one presumes, within a room inside a house that exists alongside other houses, each with their own rooms, cabinets, drawers and so on. The camera obscura of perception is here represented as a (contingent) perceptual environment that operates within, and is therefore in part articulated by, a range of architectural, social and cultural environments, rather than 'a pre-given world of objective truth'.[20]

Even Crary's claim that in Locke's *Essay* the camera obscura underwrites a self that is 'cut off from a public-exterior world' is much less straightforward than it first seems. While dividing the subject from the external world, the camera obscura of perception introduces that world (or rather the shadows of that world) deep inside the body, where the mind becomes a reader, condemned to construe from equivocal appearances pictures of the external world. Indeed, as Jules David Law observes, one 'of the central claims of empiricism . . . is that the world as we "see" it is actually flat: that what we really see when we look at the world is a two-dimensional arrangement of color, light, and figure', which the active powers of the mind, through 'inference, habit, or association . . . convert into an impression of three-dimensional space'.[21] Rather than being separate from the inside, the outside is the outside *of* the inside (and conversely, the inside is the outside of the external world). Already in *An Essay*, and then more explicitly in the eighteenth-century literature of sensibility and sensational-psychology that it influenced, perception is a meeting of inner and outer realms, which is mediated by a perceptual apparatus. '[B]y the late eighteenth century', as Geoffrey Batchen writes, 'the mirror [and we can add the camera obscura] was . . . taken as a metaphor for a dynamic enfoldment of opposites, a movement that incorporates without synthesizing the conceptual poles nature-culture, real-ideal, general-particular, science-art, object-subject, reflection-expression'.[22]

As input from the world changes (mediated, for example, by a textual, social or architectural environment), the three-dimensional space or object constructed by the subject will change. Erasmus Darwin (1731–1802) describes the visions of his textual camera obscura as an 'INCHANTED GARDEN'.[23] According to Charles Dibdin (1768–33), an inventor of spectacles and melodramas, his theatrical camera obscura shows 'the world, Sir, in little, / By light and by shadow I shew every tittle'; but he adds with heavy irony that 'Like modern philosophers . . . you always see best when you're most in the dark'.[24] Radcliffe evokes a still more radical mutation of vision when she writes in 'On the supernatural in Poetry' that at the 'potent bidding' of 'the illusions' conjured by the 'great masters of the imagination',

> the passions [are] awakened from their sleep, and . . . a crowded Theatre [is] changed to a lonely shore, to a witch's cave, to an enchanted island, to a murderer's castle, to the ramparts of an usurper, to the battle, to the midnight carousal of the camp or the tavern, to every various scene of the living world.[25]

The passions appear in this context, as they do in the second book of *An Essay*, as the product of a disjunction between idea and 'reality', and of the 'desire' this disjunction provokes (rather than emerging from a source in the body, as the humoral theory of the passions would claim). In Locke's view, desire is 'generated by a discrepancy between one's "idea of delight" and the object whose possession would provide the sensation of delight'. And desire (understood in this way) is sufficient to bring the entire spectrum of the passions into being, from joy and sorrow, to hope, fear, despair, anger, envy and shame. Indeed Locke writes that desire is 'the chief, if not only spur to human industry and action'.[26]

Locke assumes that desire will be tutored by 'reality'; but for the emergent consumer culture of the eighteenth century it suggests the possibility of novel 'ideas of delight', which will provoke 'a constant vicissitude of interesting passions'. As Anna Aikin (later Barbauld) and John Aikin write in 'On the Pleasure derived from Objects of Terror' (1773), with Walpole's *The Castle of Otranto* in mind:

> A strange and unexpected event awakens the mind, and keeps it on the stretch; and where the agency of invisible beings is introduced, of 'forms unseen and mightier far than we', our imagination, darting forth, explores with rapture the new world which is laid open to its view, and rejoices in the expansion of its powers. Passion and fancy co-operating elevate the soul to its highest pitch; and the pain of terror is lost in amazement. Hence the more wild, fanciful and extraordinary are the circumstances of a scene of horror, the more pleasure we receive from it . . .[27]

This brings us to the modern world inhabited by readers of *The Mysteries of Udolpho*, and arguably also the world explored by that novel, in which the demand for unreal-realities, and the 'new' experiences they prompt, outstrips those that can be produced by cultures still bound to the given.

Estrangement

'[I]n all the visible corporeal world, we see no chasms or gaps,'[28] Locke writes. It is accordingly reasonable to assume that the universe is an ordered whole and that, 'in this globe of earth allotted for our mansion, the all-wise Architect has suited our organs, and the bodies that are to affect them, one to another'.[29] In *The Mysteries of Udolpho* this state of harmony between subject and object, and part and whole, is represented by St Aubert (Emily's father), although it quickly comes to seem anachronistic. The death of St Aubert's infant sons leaves him without a male heir. Stripped of its past and future, his family withers. In quick succession, he learns that he has lost much of his remaining wealth; Madame St Aubert dies of a fever she catches from her husband; and St Aubert, weakened by illness and grief, is advised that to regain his health he must travel, ironically because his weakened nerves can be restored only by 'variety of scene'.[30] Given his association with stable order, it is hardly surprising that 'variety of scene' takes him to his death.

St Aubert's death sets in motion the train of events that takes Emily from La Vallée (her family home), to Toulouse, Venice, and then, in the fifth chapter of the book's second volume, to the Apennines and the villain Montoni's castle of Udolpho. Each stage of this journey takes her deeper into a world where 'chasms' divide perceptual environments and therefore observers from each other. Locke's ordered whole is consequently displaced by a proto-modernist narrative, the episodes of which are linked only by the mind that experiences them and attempts to fathom their meaning.

Emily first catches sight of Udolpho when Montoni's carriage, in which she is travelling, enters a deep valley and the 'sloping rays' of the setting sun, 'shooting through an opening of the cliffs', bring its 'towers and battlements' into view. '"There", said Montoni, "is Udolpho"', as if he were the magician who had brought it into being with the help of a natural camera obscura (sun, valley and aperture). 'Silent, lonely and sublime, it seemed to stand the sovereign of the scene, and to frown defiance on all, who dared to invade its solitary reign'.[31] The road to Udolpho ends in a place of impasse, defined by the castle's closed gates, huge portcullis, ramparts, and the 'shattered outline' of its towers. It now seems 'vast, ancient and dreary', and its towers speak of the 'ravages of war'.[32]

In the mansions described by Locke, as in La Vallée, each chamber can be located in relation to the others. Although the passage from the outside to the inside of Udolpho takes Emily through a series of darkened chambers, the drama unfolding in each chamber cannot be seen by an observer standing in any of the others, and the relation of any one chamber to the others is obscure. Inside the castle of Udolpho, space is multiple rather than singular. Rather than standing apart from events, Emily is now engulfed by them, subject to an apparently never-ending train of perceptual environments and the emotions they evoke.

After passing through the gates of Udolpho, Emily enters, in quick succession, the enclosed space of the first courtyard, sealed from its surroundings by darkness; the second courtyard, encircled by 'lofty walls, overtopt with briony, moss and nightshade, and [then by] embattled towers', which forms the hollow apex of a vast inverted pyramid; and then a grand 'gothic hall', the dimensions of which are lost in obscurity. The first leaves the impression that she is 'going into her prison'; the second brings 'long-suffering and murder . . . to her thoughts'; and the third fills her with 'timid wonder'.[33]

Next Emily enters a 'spacious apartment, whose walls, wainscoted with blacklarch-wood . . . were scarcely distinguishable from darkness itself';[34] she observes the apartment 'by the glimmer of the single lamp, placed near a large Venetian mirror, that duskily reflected the scene'.[35] This room is merely one more perceptual environment, and one more episode in the sequence we have been tracing; but it is also self-reflexive, providing a figure for Emily's and the reader's minds. In contrast to Locke's 'dark room', only a faint outline of the external world can be seen through its windows, light is provided by an internal rather than external source and the scene reflected by the Venetian mirror contains 'the tall figure of Montoni passing slowly along, his arms folded, and his countenance shaded by the plume, that waved in his hat'.[36]

Locke assumes that the understanding can stand apart from experience, which unfolds as the result of our interactions with a world framed ultimately by God. In contrast, as Emily peers into the Venetian mirror, she sees a scene within which she is immersed. Moreover, rather than being an actor in a world created by God, the scenes we have been tracing suggest that Emily is a character within a sentence or narrative composed by Montoni (the data of perception therefore provides an indirect perception of Montoni rather than God). In Udolpho, experience is no longer the stable ensemble of simple ideas imagined by Locke; instead, our simple ideas are framed by environments built and narratives composed by others, and our understanding, imagination, and passions play roles within them. This sets the stage for the still more radical revision of Lockean accounts of perception implied by the next stage of Emily's journey inside Udolpho.

After passing through a labyrinth of 'passages and galleries',[37] Emily arrives at the 'double chamber' that is to be her home in Udolpho. The chamber is 'spacious', its walls are 'lined with dark larch-wood', and during the day its windows provide a view of the surrounding countryside.[38] As its name implies, this room has two doors: the first provides access to the public spaces of the castle; the second, which can be secured only from the outside, opens to 'a steep, narrow staircase that wound from it, between two stone walls', into the castle's secret depths.[39] If we once again read allegorically, the window provides a figure for the senses, but the data they introduce is in this 'dark room' framed and potentially also articulated by the exchanges between Udolpho's public and hidden spaces. In this dangerous environment, the camera obscura of perception is itself an element within a larger perceptual apparatus, whose machinery is hardly the work of God.

Like the Earth according to Locke, Udolpho is composed of innumerable cabinets, chambers, apartments, passages, corridors, galleries and staircases. But whereas in the first, the 'phantasms' generated by our 'organs' compose a text that speaks ultimately of God, in Udolpho experience speaks only of Montoni's intentions. In La Vallée, Emily can say with Reid, 'now I yield to the direction of my senses, not from instinct only, but from confidence and trust in a faithful and beneficent monitor' who, by framing my senses, has guided me from the start.[40] But as a denizen of the second, yielding to the direction of her senses is likely to place her in the hands of Montoni.

Affective force

Jeremy Bentham devised the concept of 'sentence meaning' in order 'to capture the pressure that an ensemble of words puts on individual words'.[41] This pressure is part virtual and part actual, a result of the material arrangement of words and of the fictitious entities and imaginary non-entities that they conjure. An analogous pressure, implied by the double-chamber's second door, is exerted again and again on Emily, by the narrative into which she has been drawn. As we have seen, while Emily passes through the castle to her room, she seems a character in the process of being shaped by a narrative composed by Montoni: she passes from scene to scene, is moved by the impressions each scene leaves on her mind, and construes from these impressions an image of the terrifying world into which she is being drawn.

Once inside Udolpho, the castle's chambers, passages, galleries and staircases compose an archive of scenes which, when selected and combined by Emily, as she moves from one part of the castle to another, produces an apparently endless array of gothic fictions. Again and again Emily is 'perplexed by the numerous turnings' of the castle's passageways, uncertain how to choose from the many doors that offer themselves, and anxious that she might 'again be shocked by some mysterious spectacle'.[42] Emily has become an element – composed of selection device, dark room, lens and screen – in a perceptual apparatus able to mould her experience, rouse her passions, and so exert a shaping force on her actions and identity.

This affective force is for readers most explicitly thematized soon after Emily's arrival in Udolpho, when she writes a short note to Monsieur Quesnel (her uncle), on a blank portion of a letter Montoni has written to the same man. If, as Emily believes, Montoni's letter concerns La Vallée, then her note communicates her agreement that it be leased. But when framed by the actual content of Montoni's letter, which concerns Morano's offer to marry Emily, her note announces that she has accepted his proposal. The letter exerts a shaping pressure on the note interpolated within it, just as a sentence exerts a pressure on the words it contains, an environment on those living within it, and marriage on those it binds together.

The fiction produced by the interaction between the note and its textual environment cannot easily be dispatched; instead it rapidly becomes an element in the real world, with sufficient substantive force to counterbalance Emily's explanations of her own intentions. Montoni is left 'incredulous'. Count Morano is 'entangled in perplexity'.[43] And although Emily knows 'that neither Morano's solicitations, nor Montoni's commands had lawful power to enforce her obedience, she regarded both with a superstitious dread, that they would finally prevail'.[44]

Three chapters later this pressure takes more tangible form when Emily wakes from 'disturbed slumber' to find Count Morano advancing towards her bed. Rushing to the chamber's second door, through which Morano had entered, she discovers 'another man half-way down the steps' and, able to see 'no possibility of escape', screams 'in despair'. Leading her back 'into the chamber', Morano asks in apparently genuine confusion: '"Why all this terror? . . . Hear me, Emily: I come not to alarm you; no, by Heaven! I love you too well – too well for my own peace"'.[45]

In the fictive world of *The Mysteries of Udolpho*, Morano is actually standing before her bed, where he embodies a sexual threat. Even the narrator is relieved to report that when Emily sprang from bed she was still wearing 'the dress, which surely a kind of prophetic apprehension had prevented her, on this night, from throwing aside'.[46] Nevertheless, at this point in the narrative this threat is mediated and the danger it poses intensified by his appearance as tender lover, who has arrived to whisk her from the clutches of the villain. This shadow-show plays inside Emily's mind, drawing her by the prospect it conjures of an escape from the protean dangers of Udolpho. As the episode progresses, the shadow-show becomes more complex, evoking a world of multiple, incommensurate fictions: Montoni as protector, invoked by Emily to defend herself against Morano; Montoni as Emily's beloved, a fiction conjured by Emily's appeals to Montoni; Morano as wounded lover, whose 'darkened countenance expressed all the rage of jealousy and revenge'; Morano as villain, willing to force Emily from the castle; and admidst these phantasms, Valancourt the faithful lover, the true object of Emily's affections, who at this point in the book is in Paris, where 'his passions [have] been seduced' and 'the image of Emily' he carries in his heart has been 'obscured'.[47]

Transport

Much of the interest of *Udolpho*'s second and third volumes derives from Emily's attempts to resist the pressure exerted on her mind (and therefore her passions and actions) by the environment/narrative within which she is immersed. The consequences of transport (of being overwhelmed by affective force) are represented explicitly as marriage and implicitly as rape, robbery and murder, crimes that might be committed by any one of the brooding male figures who haunt this portion of the book: Montoni, Morano, Orsino, Cavigni, Ugo, Bertrand and so on. It is hardly surprising therefore that readers must wait until the fourth volume for an episode, the contrary of the one we have been discussing, in which a character is transported, apparently by a fictitious entity and without offering any resistance. Ludovico, the person transported, is male rather than female, a member of the servant rather than higher classes and therefore, the narrative implies, robust and insensitive enough to emerge unscathed from the experience.[48]

The events are as follows. On the north side of Chateau-le-Blanc (the white or blank Chateau), lies a long suite of rooms which lead to the 'bed-room' of the Marchioness de Villeroi, where she died in mysterious circumstances.[49] The door leading to these apartments has been locked and the apartments neglected for the past twenty years, so that the interior remains as it was left by the Marchioness. Her 'robe and several articles of her dress [are] scattered upon the chairs; her prayer-book lies open; and her bed remains with its curtains . . . half drawn'.[50] In order to quash reports that the bedroom is haunted, Ludovico volunteers to spend the night there, armed with a sword lent by his master. Late one night, he duly locks himself inside the apartments and, after ensuring that no-one else is present, enters the Marchioness's bedroom, draws his chair closer to the fire burning on the hearth, and begins to read 'The Provencal Tale'.

The plot of the tale can easily be summarized. There was once a Baron, 'famous for his magnificence and courtly hospitalities'. 'One night, having retired late from the banquet to his chamber, and dismissed his attendants', he 'was surprised [by] a stranger of noble air, but of a sorrowful and dejected countenance'. Drawing his sword, the Baron prepared to defend himself; but the Knight denied any 'hostile design'. He was carrying 'a terrible secret, which it was necessary for [the Baron] to know', but which he could divulge only if the Baron ventured alone from the castle.[51] Mastering his fears, the Baron eventually allowed himself to be led to the edge of the forest, where he found 'the body of a man, stretched at its length, and weltering in blood', that was 'the exact resemblance of . . . his conductor'. As he gazed at the corpse, the Knight vanished and

> a voice was heard to utter these words:—— . . . 'The body of Sir Bevys of Lancaster . . . lies before you. [. . .] inter the body in christian ground, and cause his murderers to be punished. As ye observe, or neglect this, shall peace and happiness, or war and misery, light upon you and your house for ever!'.[52]

This brief gothic-story establishes a complex set of parallels and echoes with the episodes we have been discussing and with *Udolpho* as a whole. The Baron, Ludovico and Emily all retire alone to a chamber where they are disturbed by a fiction, respectively: the Knight; 'The Provencal Tale'; and Morano-as-tender-lover. The first and third of these fictitious entities, mixing threat and appeal, urge their auditor to make a decision that will shape their future. Like the ghost of Hamlet's father, the ghost of Sir Bevys demands that his body be interred and 'his murderers . . . be punished'. Morano-as-tender-lover demands that Emily escape with him from Udolpho, a course of action that would put her in his power. In both cases, leaving its proper place in the realm of fantasy, fiction appears inside the everyday world. What is more, it exerts an affective force on the 'real', which threatens to transform it. Echoing Shakespeare's *Hamlet* once again, the Baron's 'future peace' depends on his response to the ghost of Sir Bevys; Emily's future happiness depends on her response to Morano-as-tender-lover.

Our third 'fictitious entity', 'The Provencal Tale', seems out of place. The relation between Ludovico and the Tale parallels that between the reader and *The Mysteries of Udolpho*, and as such seems at first to be quite different from the relation between the Baron and the ghost of Sir Bevys, Emily and Morano-as-tender-lover, and Hamlet and his father's ghost. But this difference is not as substantive as it first appears.

Like Locke's Understanding, closed up within the 'dark room' of the mind, Ludovico, the Baron, the ghost of Sir Bevys and Emily are all readers, condemned to construe from equivocal appearances (a fiction) a picture of the world. Further, 'The Provencal Tale' and *The Mysteries of Udolpho* exert a powerful affective force on their readers, not dissimilar to the force exerted by Morano-as-tender-lover and the ghost of Sir Bevys on their auditors. Echoing one of the possible outcomes of Emily's encounter with Morano-as-tender-lover, Ludovico is (in imagination) transported by 'The Provencal Tale' from the Marchioness's bedroom into the Baron's castle and then, to the edge of the forest where, after following 'an obscure and intricate path', he reaches 'a spot, where the trees crowded into a knot', and there discovers a spectacle of horror.[53]

The 'intricate path', narrative 'spot' and climactic 'spectacle' provide an analogy for the narrative of *The Mysteries of Udolpho* which, like Radcliffe's *The Sicilian Romance* (1790) and *The Romance of the Forest* (1791), was routinely praised by readers for its ability to transport readers into another world. Alan Cunningham provides a representative account in his *Biographical and Critical History of the British Literature* (1834):

> There is a fascination in her 'Mysteries of Udolpho,' which those who feel in youth will likely remember in old age: but it is not the fascination of pleasure; it resembles that practised by the adder, when it sucks, as rustic naturalists say, the lark from the sky – we shudder and become victims. The earth, as we read, seems a churchyard – the houses become castles of gloom – the streams run as if with blood – the last note of the blackbird seems that of the last trumpet – "disasters veil the moon" – and Ann Radcliffe and her mysteries triumph.[54]

'The Provencal Tale' exerts a still more profound affective force on Ludovico, which first blurs the boundary between fiction and reality and then, from the point of view of onlookers, causes it to be suspended, as Ludovico vanishes, apparently into the story he has been reading. In order to see this oscillation and the rapture with which it concludes, we must recount once more the events of 'The Provencal Tale', but this time from a position where we can see reader and character, frame and fiction, and actual and imagined worlds at the same time.

From Ludovico's point of view, 'The Provencal Tale' is fiction and *Udolpho* is reality; but as noted above, from the reader's point of view the one is linked to the other by a complex set of echoes and parallels. These parallels between part and whole frame what to the reader of *Udolpho* is an astonishing oscillation between them. After the Knight has appeared in the Baron's apartment, Ludovico thinks he can hear a noise in his own. When the Baron, now standing outside the 'great gates of his castle', thinks of 'his warm chamber, rendered cheerful by the blaze of wood', Ludovico gives 'his own fire . . . a brightening stir'. And when the Tale's narrator announces that 'a voice was heard to utter these words:——', Ludovico hears 'a voice in the chamber'.[55] He puts down the book, looks around him, but finding nothing untoward goes back to his book. At the Tale's end, Ludovico puts his book to one side, takes 'another glass of wine', and goes to sleep. But fiction now seems to have a firm hold on the real:

> In his dream he still beheld the chamber where he really was, and, once or twice, started from imperfect slumbers, imagining he saw a man's face, looking over the high back of his arm-chair. This idea had so strongly impressed him, that, when he raised his eyes, he almost expected to meet other eyes, fixed upon his own . . .[56]

In the morning, when the Count knocks on the door, there is no answer; and when the door is forced open, they find 'Ludovico's sword, his lamp, the book he had been reading', but no Ludovico. He has, it appears, been transported by a fictitious entity to another world: the reader has been articulated by the sentences of fiction rather than those of the real world.

Living machines

In Radcliffe's 'On the Supernatural in Poetry' (1826), in the course of a rather one-sided discussion between two travellers, Mr. W— and Mr. S—, the former, a literary enthusiast, argues that 'Terror and horror are so far opposite, that the first expands the soul, and awakens the faculties to a high degree of life; the other contracts, freezes, and nearly annihilates them'.[57] The distinction seems intended to justify Radcliffe's gothic fictions at the expense of those inspired by Matthew Lewis.[58] The certain horrors of the latter stop readers in their tracks, appalled by what has been described. The uncertain terrors of the former make readers, along with the book's characters, actively involved in the struggle to discern the truth behind equivocal appearances. They therefore become co-creators of imagined worlds built from the incomplete materials provided by the text.

It is usually argued that imagined worlds must ultimately be brought into determinate relation with reality. The second traveller, Mr S—, introduced as one 'who seldom troubled himself to think upon any subject, except that of a good dinner', echoes the views of St Aubert, Montoni and Laurentini when he warns his friend that '"an object often flatters and charms at a distance, which vanishes into nothing as we approach it; and 'tis well if it leave only disappointment in our hearts. Sometimes a severer monitor is left there"'.[59]

At first glance the book follows a similar path, although to a more positive conclusion, by revealing that behind the veil of appearances is only mundane reality. In the last pages of the book we learn, for example, that Ludovico has been transported by villains rather than ghosts, who have gained access to the Marchioness's bedroom through a secret passage that leads to a hidden door in the wall. But when Radcliffe writes that obscurity 'awakens the faculties to a high degree of life', she does not go on to say that as a result we are able to find what is behind the veil. The primary effect of Udolpho's anti-climactic conclusion and explained supernatural is to reveal just how active Emily's and the reader's imaginations have been, by bringing reader and heroine to a point where they can see themselves seeing, see themselves reading, and watch their own passage through the book. But rather than revealing the reality behind the fiction, this revelation prepares the ground only for another fiction, generated by yet another imagination machine.

We can therefore say that, in the absence of unmediated reality, Radcliffe's explained supernatural draws our attention to both the active and passive poles of perception. On the one hand, it foregrounds the recognition, already evident in the empiricist philosophies of Locke and Hume, that perception proceeds as if it were the work of a machine. As Coleridge implies and Radcliffe's novels demonstrate again and again, perceptual environments (whether textual, architectural or social) working in tandem with the camera obscura of perception, are productive of *affect* (rather than the inert 'simple ideas' described by Locke). Emily must learn to manage the metamorphic force that perceptual environments and the fictitious entities they conjure exert on her emotions and, therefore, on her actions and identity. On the other hand, the explained supernatural draws attention to the role played by Emily and the reader in constructing events. In a dynamic closely entwined with the first, affect becomes expression as Emily attempts to construe the world implied by circumstances, and as she interprets and renarrates her disparate experiences, often by attempting to draw them into a sequence in which they might make sense.

The tension between affect and expression marks Emily as a modern subject whose identity is based on desire rather than humoral passion or innate ideas, which is to say, following Locke, on the 'discrepancy between one's "idea of delight" and the object whose possession would provide the sensation of delight'. It informs the episodes and drives the narrative of *The Mysteries of Udolpho*. Indeed, when this tension is resolved (in the marriage feast held to celebrate the marriage of Emily and Valancourt) the novel comes to a close.

In contrast to Radcliffe, whose novels hold these poles in precarious balance, Coleridge routinely foregrounds expression at the expense of affection, and the creative rather than machinic elements of perception, by troping the productive and metamorphic power of fiction as evidence of life's creative power, as exemplified by the imagination of the poet. We are reduced to 'a passive Automaton, or *mere* Camera Obscura' (my italics), when our minds become screens for the moving phantasms of consumer culture. But when our faculties wake from sleep, expression eclipses affect. Animated by this living power, painting (and perception as well) become creation rather than re-presentation, 'or if painting, yet such, and with such co-presence of the whole picture flash'd at once upon the eye, as the sun paints in a camera obscura'.[60] In contrast to gothic fictions, the 'power of Poetry,' Coleridge writes, 'is by a single word to produce that energy in the mind as compels the imagination to produce the picture'.[61]

The opposition drawn by Coleridge, between the camera obscura of imagination and the *mere* camera obscura, expression and affect, living beings and automata, is compelling. Yet the presence of the camera obscura at both poles underlines just how difficult it is for the poet's creations to divide themselves from their machinic doubles (the reader, printer, media and market on which they depend). At the same time, the proliferation in modern consumer cultures of *mere* camera obscuras (able to deliver *ab intra* delusions as if they were *ab extra* projections) ensures that the poet's 'living' creations will always be outnumbered by, and often indistinguishable from, their 'lifeless' counterparts. In this context, the path we have traced, from estrangement to transport, defines the path of a sublime of consumption that concludes with the search for novel means of transport, rather than a revelation of the ultimate nature of things.

Coleridge's negative judgement of gothic fictions turns out to be the story of his own gothic nightmare, described most vividly in 'The Pains of Sleep', *and* of the ungrounded transports celebrated by a consumer culture in which the demand for unreal-realities, and for the 'new' experiences they prompt, outstrips those that can be produced by cultures still bound to the given. The temptation to conclude by adjudicating between these competing stories is almost irresistible; and yet both are necessary if one is properly to understand the roots of the still tortured agon between the demands of mind and machine, the human and the automaton, and expression and affective force.

Notes

1. Samuel Taylor Coleridge, *Biographia Literaria*, ed. J. Engell and W. Jackson Bate, 2 vols, vol. vii of *The Collected Works of Samuel Taylor Coleridge*,

16 vols, Bollingen Series, 75 (London: Routledge and Kegan Paul; Princeton: Princeton University Press, 1983), i: 48n. The following argument draws in part on the account of *The Mysteries of Udolpho* advanced in Chapter 4 of my *Multiplying Worlds: Romanticism, Modernity, and the Emergence of Virtual Reality* (Oxford: Oxford University Press, 2011).

2. Horace Walpole, 'Preface to the First Edition', in W.S. Lewis (ed.), *The Castle of Otranto: A Gothic Story* (Oxford: Oxford University Press, 1986), 3–6: 4.
3. Ann Radcliffe, *The Mysteries of Udolpho*, World's Classics (1980; rpt. Oxford: Oxford University Press, 1988).
4. Mark Pendergrast, *Mirror/Mirror: A History of the Human Love Affair with Reflection* (New York: Basic Books, 2003), 85.
5. Pendergrast, *Mirror/Mirror*, 85.
6. Isaac Newton, *Opticks or A Treatise of the Reflections, Refractions, Inflections and Colours of Light*, 4th edn (1730; rpt. New York: Dover, 1952), 15. See also René Descartes, *The Philosophical Writings of Descartes*, trans. John Cottingham, Robert Stoothoff and Dugald Murdoch, 3 vols (Cambridge: Cambridge University Press, 1984–91), ii:166.
7. John Locke, *An Essay Concerning Human Understanding*, ed. Alexander Campbell Fraser, 2 vols (1894; rpt. New York: Dover, 1959), i: 212 (Book II. chapter xi. paragraph 17).
8. Locke, *An Essay*, i, 121 (II. i. 2). For background to the controversy over innate ideas see: Stephen P. Stich (ed.), *Innate Ideas* (Berkeley: University of California Press, 1975).
9. Nancy Armstrong and Leonard Tennenhouse, 'A Mind for Passion: Locke and Hutcheson on Desire', in Victoria Kahn, Neil Saccamano, and Daniela Coli (eds), *Politics and the Passions, 1500–1850* (Princeton: Princeton University Press, 2006), 131–50: 134.
10. Armstrong and Tennenhouse, 'A Mind for Passion', 134.
11. Jonathan Crary, *Techniques of the Observer: On Vision and Modernity in the Nineteenth Century* (Cambridge, MA: MIT Press, 1992), 38–40.
12. Crary, *Techniques of the Observer*, 24.
13. Jonathan Crary, 'Modernizing Vision', in Hal Foster (ed.), *Vision and Visuality*, Dia Art Foundation Discussions in Contemporary Culture, no. 2 (New York: The New Press, 1988), 6, 34.
14. Locke, *An Essay*, i, 122–3 (II. i. 3). Locke nevertheless assumes the existence of a real world lying behind and mirroring the perceived one.
15. Locke, *An Essay*, i, 169 (II. viii. 8).
16. David Hume, *A Treatise of Human Nature*, ed. Ernest C. Mossner (1739–40; London: Penguin, 1985), 132. Locke admits that whether creations of the imagination, such as a centaur, 'can possibly exist or no, it is probable we do not know', *An Essay*, i: 501 (II. xxx. 5).
17. Thomas Reid, *An Inquiry into The Human Mind, on the Principles of Common Sense* (Edinburgh: Printed for A. Millar, London, and A. Kincaid & J. Bell, Edinburgh, 1764), 119.
18. Locke, *An Essay*, i, 146 (II. ii. 3).
19. Locke, *An Essay*, i, 48 ('Introduction'. i. 15).
20. Crary, *Techniques of the Observer*, 39–40.
21. Jules David Law, *The Rhetoric of Empiricism: Language and Perception from Locke to I. A. Richards* (Ithaca: Cornell University Press, 1993), 3–4.

22. Geoffrey Batchen, *Burning with Desire: The Conception of Photography* (Cambridge, MA: MIT Press, 1997), 81.
23. Erasmus Darwin, *The Loves of the Plants* (1789; rpt. Oxford: Woodstock Books, 1991), v–vi.
24. Charles Dibdin, 'The Camera Obscura', in *Mirth and Metre, consisting of Poems, Serious, Humorous, and Satirical; Songs, Sonnets, Ballads & Bagatelles* (London: Vernor, Hood, and Sharpe, 1807), 144–6: 145.
25. Ann Radcliffe, 'On the Supernatural in Poetry', in E.J. Clery and Robert Miles (eds), *Gothic Documents: A Sourcebook 1700–1820* (Manchester and New York: Manchester University Press, 2000), 163–72: 169. First published in *New Monthly Magazine*, 16: 1 (1826), 145–52.
26. Locke, *An Essay*, i. 304 (II. xx. 6).
27. John and Anna Letitia Aikin, 'On the Pleasure derived from Objects of Terror; with Sir Bertrand, a Fragment', in *Miscellaneous Pieces, in Prose* (London, 1773), 117–37: 125.
28. Locke, *An Essay*, ii, 67 (III. vi. 12).
29. Locke, *An Essay*, i, 402–3 (II. xxiii. 12).
30. Radcliffe, *The Mysteries of Udolpho*, 25.
31. Radcliffe, *The Mysteries of Udolpho*, 226–7.
32. Radcliffe, *The Mysteries of Udolpho*, 227.
33. Radcliffe, *The Mysteries of Udolpho*, 227–8.
34. Radcliffe, *The Mysteries of Udolpho*, 228.
35. Radcliffe, *The Mysteries of Udolpho*, 229.
36. Radcliffe, *The Mysteries of Udolpho*, 229.
37. Radcliffe, *The Mysteries of Udolpho*, 232.
38. Radcliffe, *The Mysteries of Udolpho*, 234.
39. Radcliffe, *The Mysteries of Udolpho*, 235.
40. Reid, *An Inquiry*, 415.
41. Frances Ferguson, *Pornography, the Theory: What Utilitarianism Did to Action* (Chicago and London: University of Chicago Press, 2004), 3. See also John Bender, *Imagining the Penitentiary: Fiction and the Architecture of Mind in Eighteenth-Century England* (Chicago: University of Chicago Press, 1987), 214–15, and Ross Harrison, *Bentham* (London: Routledge and Kegan Paul, 1983), 59.
42. Radcliffe, *The Mysteries of Udolpho*, 258.
43. Radcliffe, *The Mysteries of Udolpho*, 202.
44. Radcliffe, *The Mysteries of Udolpho*, 209.
45. Radcliffe, *The Mysteries of Udolpho*, 260–1.
46. Radcliffe, *The Mysteries of Udolpho*, 261.
47. Radcliffe, *The Mysteries of Udolpho*, 652.
48. The events introducing this episode explicitly link Ludovico's experiences with Emily's in the castle of Udolpho.
49. Radcliffe, *The Mysteries of Udolpho*, 547.
50. Radcliffe, *The Mysteries of Udolpho*, 533, 532.
51. Radcliffe, *The Mysteries of Udolpho*, 552–3.
52. Radcliffe, *The Mysteries of Udolpho*, 556.
53. Radcliffe, *The Mysteries of Udolpho*, 556.
54. Allan Cunningham, *A Biographical and Critical History of the British Literature of the Last Fifty Years* (Paris: Baudry's Foreign Library, 1834), 124.

55. Radcliffe, *The Mysteries of Udolpho*, 555–56.
56. Radcliffe, *The Mysteries of Udolpho*, 557.
57. Ann Radcliffe, 'On the Supernatural in Poetry', 168.
58. See for example Matthew Lewis, *The Monk: A Romance* (Waterford: Printed for J. Saunders, 1796) and the numerous writers who followed in his footsteps with tales of graphic violence, sexual transgression and supernatural terrors.
59. Radcliffe, 'On the Supernatural in Poetry', 163.
60. Coleridge, *Biographia Literaria*, i. 128. Coleridge is referring to John Milton's *Paradise Lost*, Book IX, lines 1101–10.
61. Samuel Taylor Coleridge, *Lectures, 1808–1819, on Literature*, ed. R.A. Folkes, vol. v of *The Collected Works of Samuel Taylor Coleridge*, 16 vols, Bollingen Series, 75 (London: Routledge and Kegan Paul; Princeton: Princeton University Press, 1983), 523.

2
Air-Looms and Influencing Machines

Steven Connor

Just over two centuries ago, just as the machine began to impinge most closely upon human consciousness, began to twin with it, machinery went mad. Or, what may come to the same thing, the mad became mechanical. They began to dream of machines, complicated dreams of sinister, far-reaching apparatuses. The mad machinery of these dreams is not a machinery out of control – not like the juridical machine of Kafka's *In the Penal Settlement*, which ends by systematically disassembling itself. It is a machinery that is mad in its functioning, its madness the madness of its very reliability, its repeatability. For the essential madness of the machine lies in its apparent rationality, the offer it holds out to rationality to find its like in mechanical process. For a sane man to believe seriously that he has become a machine is to become mad. A hundred years after what looks like its first appearance, Viktor Tausk would give this machine a name, suggesting that all the machineries of the mad are varieties of or modular components of the mysterious, polymorphous apparatus he calls the 'influencing machine'. The purpose of this machine, which can be operated by persecuting entities over large distances, is to exert control over the patient's thoughts and feelings. Tausk explains that the influencing machine typically makes the patient see pictures (and when it does this, it takes the form of a magic lantern or cinematograph); it 'produces, as well as removes, thoughts and feelings by means of waves or rays or mysterious forces'; it produces motor phenomena in the body, like erections and emissions, by means of 'air-currents, electricity, magnetism, or X-rays'; it creates indescribable bodily sensations, 'that in part are sensed as electrical, magnetic, or due to air-currents'; and brings about other bodily symptoms such as abscesses and skin-eruptions.[1] But the most important function of the machine is its systematic dealing out of delusion, including, we must suppose, the master-delusion, of the influencing machine itself. For the maddest thing about this mad contraption is that it is its own product, that its purpose seems to be to *manufacture, maintain and ramify itself.* Tausk makes it clear that the influencing machine always has a history, its workings becoming ever more complex and elaborate as the patient's psychosis becomes more deeply rooted. This is because the machine is also the patient's strongest defence against his madness, in providing a model and an explanation for how his madness works. The maddest thing about this machine is the knowledge it seems to provide of the workings of the madness, which is to say, of itself. Nowhere are the melancholy-mechanical mad madder than in the image they have and hold to of their madness.

Psychoanalysis and the varieties of corporealized psychiatry similarly propose more or less mechanical systems as correlatives for the act of thinking. In fact, the influencing machine can be an exquisite parody of the psychoanalytic process itself, especially its tendency to convert the spiritual entities of previous eras (bad spirits and possessing demons) into impersonal forces and agencies – id, ego, superego, repression, cathexis, libido. There is thus a strange concurrence between illness and remedy in the case of mechanical delusions, or delusive machines. Tausk reports a number of cases in which patients felt flows and currents, including that of a man who 'felt electrical currents streaming through him, which entered the earth through his legs; he produced the current within himself, declaring with pride that that was his power!' But Tausk matches this conception in his explication of the processes behind the formation of the influencing machine, which requires us to accept the assumption 'that the libido flows through the entire body, perhaps like a substance (Freud's view), and that the integration of the organism is effected by a libido tonus, the oscillations of which correspond to the oscillations of psychic narcissism and object libido'.[2] This flow of libido is far from merely metaphorical, since it is capable of bringing about transient swelling of organs, as a result of 'an overflow of secretion resulting from libidinal charging of organs'.[3] Both madness and its analysis are explications of the workings of machinery: of the machinery of delusion in the case of the physician and of the delusive machine of the patient, by which latter is meant the machine produced by delusion and the machine that deals out delusions, especially the delusion of itself.

Readers of the elaborate delusions of Daniel Paul Schreber in particular have increasingly drawn attention to the importance of technologies of communication – telegraphs, telephones, radio – in his version of the influencing machine. The tendency has been to read all influencing machines as anticipations of the paranoia associated with such distinctively modern technologies of communication. I want to focus attention on a particular feature of the physical workings of the influencing machine, namely the ways in which they are held to work over space and, more specifically, how they imagine the mechanical manipulation of what fills that space – whether air, ether or some other insubstantial matter. Conjugating a number of systematic accounts of systematic delusions, those in particular of James Tilly Matthews, Friedrich Krauß and John Perceval, I will try to construe the workings of the influencing machine as a pneumatics, or a fluid mechanics.

Gaz-plucking

In the Autumn of 1809, a writ for habeas corpus was delivered to Bethlem Hospital ('Bedlam') demanding the release of an inmate who had been confined in the hospital for 13 years. This led to a hearing before the King's Bench in which affidavits from the family of the confined lunatic were considered against those furnished by Bethlem Hospital itself. Chief among the affidavits on the side of the confined man was an account by two doctors, one George Birkbeck MD and his associate Henry Clutterbuck, of their visits to the prisoner over the course of some months, leading to their conclusion that he exhibited no traces of insanity and should be released. Against them, the governors of the hospital brought forward various witnesses as to the confined man's dangerous insanity, which extended to making threats against the King and his ministers. Most tellingly of all, they quoted a letter of 7 September 1809 from no less a person than Lord Liverpool, the Home Secretary, recommending that the prisoner continue to be entertained at the public expense in Bethlem Hospital. The case for habeas corpus was rejected.

It was not usual for the Home Secretary to be applied to in such a case. It seems likely that it was not in his office but in his person that Lord Liverpool was addressed. For this particular lunatic had first been brought to Bedlam following an outburst he had made on 30 December 1796 from the public gallery of the House of Commons during a speech made by Robert Banks Jenkinson, Lord Liverpool himself, defending the government against the charge that it had deliberately obstructed efforts to avert the costly and damaging war against France. Lord Liverpool's speech had been interrupted by a cry of 'Traitor!' from the gallery and a man had been hustled away by stewards. During his examination by the Bow Street Magistrates, it emerged that the prisoner's name was James Tilly Matthews, who had a wife and young son. Matthews had been a tea merchant, but claimed to have been in the service of the government since just before the outbreak of war against France, acting as an emissary conducting secret peace talks with the revolutionary government in France. Now the government had decided against peace, seeing their opportunity of crushing decisively the political threat from across the Channel and halting the drift of revolution. Matthews' sense of betrayal had led to his public denunciation of Lord Liverpool, the minister whom he most particularly blamed for his predicament. The fierce threats that Matthews uttered against Lord Liverpool and other members of the government, combined with his obviously excitable condition, left the Bow Street magistrates with no choice but to request the governors of Bethlem Hospital to take custody of him.

The story of what led to the outburst as a result of which James Tilly Matthews was confined in Bethlem Hospital, and why Lord Liverpool should have continued to take an interest in him has been told a number of times, most recently and in the most fluent and illuminating detail by Mike Jay.[4] The most important thing about the 1810 habeas corpus hearing was the effect it had on John Haslam, the resident apothecary and senior medical officer of the Bethlem Hospital, spurring him into publishing an elaborate defence of his view that Matthews was completely insane and unfit to be released. Stung by the challenge to his professional judgement, Haslam was determined to show that Matthews was not only mad, but epically and self-evidently so. The way to do this was, as he put it 'to develop the peculiar opinions of Mr. Matthews, and leave the reader to exercise his own judgement concerning them'.[5] His *Illustrations of Madness* accordingly gives Matthews' delusions the fullest possible airing, often using his own words, very likely taken down by Haslam himself, and giving a richness of detail unprecedented in the literature of madness. The result is a kind of accidental phenomenology, an insider account of Matthews' mad-machine and machine-madness.

Haslam explained that Matthews believed himself to be subject to violent and continuous persecution by a gang of 'villains profoundly skilled in Pneumatic Chemistry', who operated from a basement not far from the hospital and assailed him by means of an apparatus called an 'Air Loom'. Haslam gives Matthews' roster of the seven members of the gang: 'Bill the King', their mysterious leader; 'Jack the Schoolmaster', a shorthand writer who records all the gang's doings; Sir Archy, a foul-mouthed blackguard, who is possibly a woman dressed as a man; the

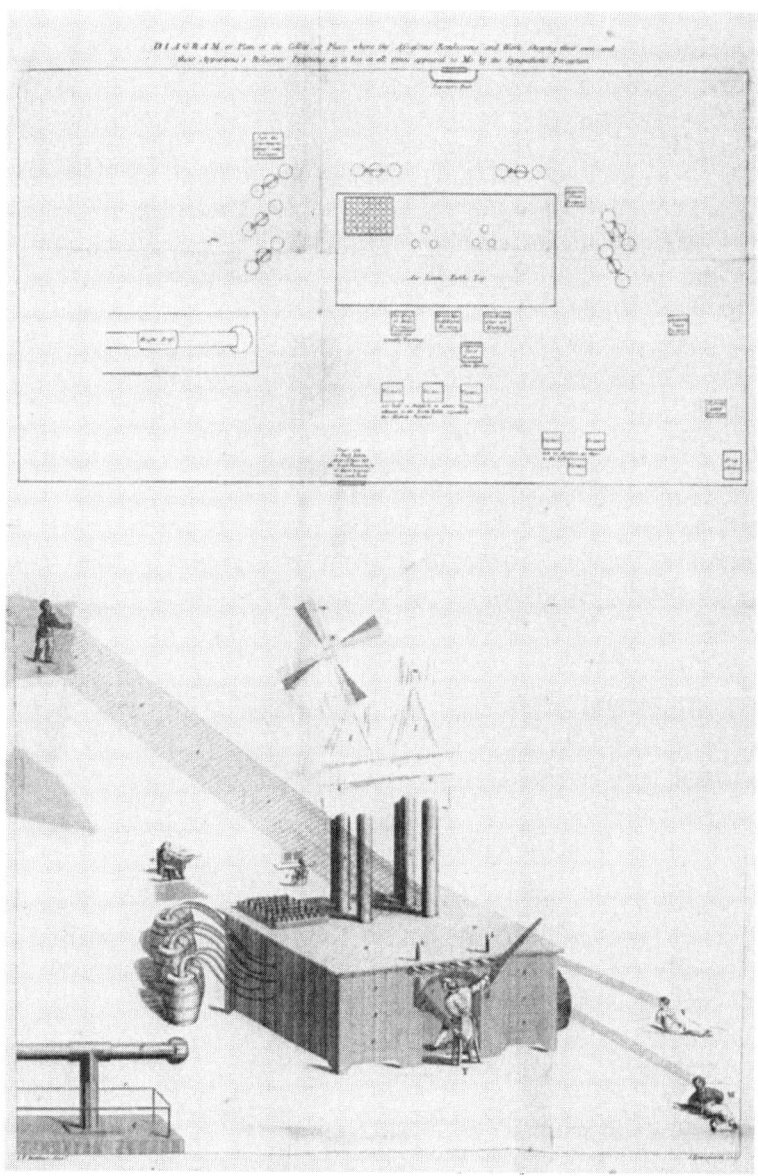

Figure 3: James Tilly Matthews' 'Air-Loom', from John Haslam's *Illustrations of Madness* (1810)

'Middle Man', the most skilful operator of the instrument; 'Augusta', a 36-year-old woman, who is frequently out and about; 'Charlotte', apparently French, and kept a prisoner by the gang; and finally one who has no name but the 'Glove Woman', who keeps her arms covered because of the itch, operates the machine with skill, but has never been known to speak. Another document of Matthews, not quoted by Haslam, but included in Roy Porter's edition of *Illustrations of Madness*, suggests that there may be more even than this, for he speaks of '[t]he dreadful gang of 13 or 14 Monster Men & Women who are so making their Efforts on Me'.[6]

The most extraordinary feature of Haslam's report is the drawing, by Matthews himself, which he provides of the Air-Loom, along with Matthews's own detailed key to its elements. The machine seems to have worked in something like the following fashion. Noxious vapours contained in barrels are directed through pipes into the principal body of the apparatus, which is contained invisibly within an enormous desk-like structure (one of a number of features in the machine that suggests the parallels between the machine and the psychic bureaucracy that Daniel Paul Schreber would call an 'Aufschreibesytem', a 'writing-down-system'). As a result of some mysterious process of distillation occurring inside the machine, a magnetic-mesmeric fluid is produced, and perhaps stored in the battery-like 'cluster of upright open tubes or cylinders, and, by the assassins, termed their *musical glasses*'.[7] These objects may also be meant to be magnets, to which Matthews refers, and which do not seem otherwise to be visible. It is drawn off through the tubes emerging from the apparatus and transmitted by a 'windmill kind of sails'.[8] The substance itself is described as

> The warp of magnetic-fluid, reaching between the person impregnated with such fluid, and the air-loom magnets to which it is prepared; which being a multiplicity of fine wires of fluid, forms the sympathy, streams of attraction, repulsion, &c. as putting the different poles of the common magnet to objects operates; and by which sympathetic warp the assailed object is affected at pleasure.[9]

Edmund Cartwright had invented a power-loom that could be operated by steam or water in 1785, but Matthews's machine still appears to be largely hand-operated, and its operation seems to require considerable finesse on the part of the gang. Matthews draws attention to the levers, 'by the management of which the assailed is wrenched, stagnated, and the sudden-death efforts made upon him'.[10] He also refers to '[t]hings, apparently pedals, worked by the feet of the pneumaticians' and '[s]omething like piano-forte keys, which open the tube valves within the air-loom, to spread or feed the warp of magnetic fluid', though this last does not appear in Matthews' illustration.[11]

The effects of the fluid on its victim are exotic and fearsome. They include: *'Fluid-locking'*, a constriction of the tongue which prevents speech; *'Cutting soul from sense'*, in which the heart and the intellect are separated by a veil of magnetic fluid; *'Stone-making'*, the formation of calculus in the bladder; and *'Thigh-talking'*, in which the victim becomes convinced that 'his organ of hearing, with all its sensibility' is lodged on the external portion of the thigh, to which utterances are magnetically directed.[12]

Matthews is also subject to the violently kinetic effects of pneumatics. The most fiendish torture of all (the one in which the operator in the illustration is in fact engaged), is called *'Sudden-death-squeezing'* or *'Lobster-cracking'*, described as 'an external pressure of the magnetic atmosphere surrounding the person assailed, so as to stagnate his circulation, impede his vital motions, and produce instant death'.[13] There is also *'Laugh-making'*, which 'consists in forcing the magnetic fluid, rarified and subtilized, on the vitals [*vital touching*] so that the muscles of the face become screwed into a laugh or grin'.[14] Not only is he subject to impregnation with the magnetic fluid, the operators make him a pharmaceutical factory for the production of more fluid:

> *Gaz-plucking* is the extraction of magnetic fluid from a person assailed, such fluid having been rarified and sublimed by its continuance in the stomach and intestines. This gaz is in great request, and considered as the most valuable for the infernal purposes of these wretches. They contrive, in a very dexterous manner, to extract it from the anus of the person assailed, by the suction of the air-loom. This process is performed in a very gradual way, bubble by bubble.[15]

We are left to imagine for ourselves the exquisite tortures encompassed in the terms *'foot-curving, lethargy-making, spark-exploding, knee-nailing, burning out, eye-screwing, sight-stopping, roof-stringing, vital-tearing* [and] *fibre-ripping'*.[16]

Mind, matter, metaphor

But, for all the horrifyingly corporeal tortures it inflicts, the air-loom is much more intelligible as a reflection on Matthews' mind, or on the processes of thought, though indeed the result is a dramatically corpo-realized thinking. The air-loom is an allegory of the process of thought, perhaps an allegory of the process of imaging, of allegorizing thought, of making matter of it.

There is another feature of the air-loom, apparent not only in Matthews' but also in other such systematic delusions, that is rarely discussed, though all readers from Haslam onwards are surely aware of it, namely the fact that the whole apparatus, the whole conception of the air-loom, is so richly and deliciously risible. Matthews must surely have been aware of the arguments of magnetizers and mesmerists that the vital fluid which they were manipulating was supposed to derive its power from its ethereal fineness. And yet this superfine stuff is supposed to be distilled from the scavenged stinks contained in the barrels which decant into the machine, which are much more the kind of thing one would expect to find in a witch's cauldron than in the workshop of a pneumatic technician:

Seminal fluid, male and female – Effluvia of copper – ditto of sulphur – the vapours of vitriol and aqua fortis – ditto of nightshade and hellebore – effluvia of dogs – stinking human breath – putrid effluvia – ditto of mortification and of the plague – stench of the sesspool – gaz from the anus of the horse – human gaz – gaz of the horse's greasy heels – Egyptian snuff (this is a dusty vapour, extremely nauseous, but its composition has not been hitherto ascertained) – vapour and effluvia of arsenic – poison of toad – otto of roses and of carnation.[17]

Though Matthews stresses the fact that the machine is built on the basis of the most fearsomely up-to-date science, what strikes us most mark-edly in its representation is the grotesque collage of the ancient and the modern, the efficient and the laborious, the cerebral and the corporeal. A machine is a reproduction and often also a grotesque parody of living process (hence the offensiveness of representing living process as 'mere' mechanicity). Matthews seems to give us a parody of the very idea of a machine, which seems to draw attention to its own status as sorry fantasy. It is almost as if Matthews were giving us the very image of an obviously spoof machine, the purpose of which is to demonstrate its own implausibility.

The conspicuous comedy of Matthews' machine derives in part from the Bergsonian recoil of the *élan vital* from the idea of human thought being subjugated by a series of gimcrack contrivances like this one. Curiously, though, there seems to be a kind of meeting of minds between Matthews and Haslam on this score. Both of them have a tendency to view the operations of the mind in bizarrely material terms, and they seem to agree that the disordering of sense and idea characteristic of madness is a result of strictly physical causes. As one might perhaps expect from the superintendent of an asylum, Haslam does not have a very high opinion of the powers of the human mind.

> As far as I have observed respecting the human mind (and I speak with great hesitation and diffidence,) it does not possess all those powers and faculties with which the pride of man has thought proper to invest it. By our senses, we are enabled to become acquainted with objects, and we are capable of recollecting them in a great or less degree; the rest, appears to be merely a contrivance of language.[18]

The limitation of the mind's powers is nowhere more painfully apparent than when it tries to reflect upon its own constitution or operations:

> If mind, were actually capable of the operations attributed to it, and possessed these powers, it would necessarily have been able to create a language expressive of these powers and operations. But the fact is otherwise. The language, which characterizes mind and its operations, has been borrowed from external objects; for mind has no language peculiar to itself.[19]

'[W]e live in a world of metaphor,' Haslam confirms in a later footnote.[20] It is for this reason that he has so little faith in reasoning with the mad, or dealing with them on their own terms, for this would mean giving credence to the delusions from which they need to be delivered. This is particularly the case with 'methodical madness' like that of Matthews (though Matthews does not seem to be mentioned or referred to in this earlier text), since '[i]n proportion as insanity has assumed a systematic character, it become [sic] more difficult of cure'.[21] Haslam does not believe there can be such a thing as a disease of the mind, since the mind is incorporeal. However, a 'disease of ideas', caused by a disordering of the material in which mind is embedded (and from which it may be required to derive its ideas of itself) is conceivable, though also deeply intractable.

Ingenious engines of thought appear in the delusional systems of other nineteenth-century madmen. None of them seem as fully developed or autonomous as Matthews', nor do they seem to have been visualized as clearly and continuously as the air-loom. They function as machinery in a more general sense, rather than as single machines. But perhaps they can nevertheless be seen as adjuncts or outworks to Matthews' machinery. And all of these machineries have a pneumatic component, though it may be less obvious or central than in Matthews'.

Life-Ether

The first of Matthews' heirs of whom we know was Friedrich Krauß. Born in 1791 in Göppingen, after a distinguished academic career Krauß was appointed to the firm of Daniel Thuret and sent to Antwerp in 1814, where his delusions of persecution began in 1816. After a period of confinement in the Cellitenkloster Institution in Antwerp, he was returned home to Göppingen. Here, he wrote the first of his many petitions to the German and Belgian authorities, begging them to take measures against his persecutors. From 1819 to 1824, he gave classes in Heidelberg in languages, commercial law and technical chemistry. From 1827, he resumed his occupation as a travelling commercial representative. In 1832, he began keeping a diary in which he detailed his various mental torments. These were to last his entire life, and to form the subject of two long autobiographical accounts, *Nothschrei eines Magnetisch-Vergifteten* (1852) and *Nothgedrungene Fortsetzung meines Nothschrei* (1867), written only some years before his death. There is only one surviving copy of the first volume, in the library of the University of Tübingen. The library of the Society for Psychical Research holds the only surviving copy of the latter. Only selections from the texts have been made available in modern editions.[22]

Krauß wrote of his torment by a number of voices, which soon coagulated into characters. During the early phase of his persecution, he tried to kill himself by dashing his head against a wall; but, as in other such cases, the more the agencies became persons, the easier it seems to have become for him to cope with them, as his anxiety turned into outrage.[23]

There were four principal persecutors, whom he called The Old Magnetizer, Janeke Simon-Thomas, Van Asten and the daughter of Van Asten. They could read his thoughts (leading him to call them 'soul-worms' and 'thought-vultures').[24] They also exercised control over his feelings and moods, producing depression and numbness, as well as distorting his senses of hearing and touch and inflicting actual physical torments. At times, he felt distinctly that his vocal apparatus had been completely taken over and directed according to the will of his persecutors.[25] During all this time, Krauß struggled to maintain his profession and social position, even though his tormentors would constrain him to perform violent and inappropriate actions, such as leaps into the air, while he was trying to conduct business.

Since he remained uninstitutionalized, Krauß was free to seek information regarding the theories of magnetism and mesmerism as well as professional advice from scientists and physicians, some of whom confirmed his belief that he was subject to magnetic influence, but were unable to assist him. He believed that the publication of his work would encourage others subject to the same torments to come forward, and suggests that doctors told him of other cases known to them of persecution by animal magnetism.[26] He seems to have known of Matthews' account of his illness, if only through hearsay or repute.[27] Thirty pages of the 1852 *Nothschrei* were given over to correspondence between Krauß and a Herr L'Hermet of Magdeburg, who also believed himself to be magnetized. L'Hermet had managed to keep his auditory hallucinations under control, so that he heard at most 'a light hissing or sizzling', though bodily hallucinations continued strongly, for example of 'numbing ether-stuff pouring into the ears'.[28]

Krauß believed that the material means whereby his persecution was affected was magnetism, which, like many others, he imagined as a form of electricity. Later he would come to refer to it as 'Life-Ether', 'Lebensäther'.[29] Although animal magnetism resembles electricity, it has an advantage over it:

> As is well known, electricity is transmitted only as long as there is in contact with the wire or with the person who is holding it. By contrast, the electromagnetic fluid maintains its course like a fishing line through the body affected by the magnetiser; once that body has been attracted by the discharge of the apparatus, a flow of influence is guaranteed through all circumstances, and for as long as the magnetiser wishes.[30]

Krauß too believed himself to be subject to ear-entry by the malignant gases. He explains that '[t]he ears are the channels to which Nature has given the office of . . . drawing out the magnetic fluid from the air of the world, taking it in and introducing it into the body . . . It is for this reason that this concentrated ether entered me through the ears'.[31]

Krauß believed that the magnetizers used various carriers or media to transmit the fluid, but in particular gases. At one point he distinguishes carefully between three different forms of magnetic gas:

> From the beginning, I distinguished three . . . distinct types of mag-
> netic gas: 1) the usual, that flowed in with only a weak sound, like
> the hissing of boiling water 2) that which steamed in with a loud siz-
> zling and grating, like the scraping of sand, and powerfully touched,
> stretched and suffused nerves, muscles and veins . . . 3) the most
> concentrated, densest and most powerful gas. This was drawn out in
> a very high-pitched sound, like hoooo or tsiiiii . . . like a living flame
> that reached in terribly, stimulated everything to its highest pitch,
> kindled everything in an instant and caused me the greatest pain,
> the most fearsome agony.[32]

Krauß suggested that the magnetizers used machines in order to transmit their streams of magnetic influence to him. Though he has clearly
read widely in the theories of animal magnetism, citing in particular
D.G. Kieser's *System des Tellurismus* of 1826,[33] and has plenty to say
about them, he is much less interested than Matthews in determining
or explicating the nature of these machines.

It is difficult to be sure whether Krauß's machine is less developed
than Matthews' or more diffused. Krauß emphasizes that the secrets of
electromagnetism are spreading ever wider, and taking ever more various forms. The widespread use of electricity meant that 'what was once
thought to be a secret concealed by a few specialists and so-called black
magicians, now lies so open to view, that the multifarious applications
of this power are becoming ever more general'.[34]

> [t]his animal or electromagnetism, also known as tellurism, as the
> mightiest ethereal power, as Life-Ether, has immeasurable reach;
> and, through the continued researches of brilliant intellects, has
> expanded into the sphere of the incredible, and may easily bring
> about a transformation in spiritual matters greater than that effected
> by steam in material life.[35]

Animum rege

John Perceval was the son of Spencer Perceval, the Prime Minister who
was assassinated in the lobby of the House of Commons in May 1812,

when his son was aged nine. John Perceval joined a cavalry regiment, and, after seeing service in Portugal, rose to the rank of Captain. In 1830, he left the army and travelled to Scotland, where there was an outbreak of charismatic Christianity, characterized by speaking in tongues and other ecstatic religious appearances. Already, Perceval was beginning to behave in strange ways, finding himself impelled – or allowed – to break into religious utterance. He travelled from Scotland to Dublin, where his behaviour became so very disordered that he had to be restrained in a room of the inn where he was staying. In December 1830, his brother came to fetch him, and took him to an asylum run by a Dr Fox near Bristol. He was confined in this institution for eighteen months, and then transferred to a second institution, at Ticehurst in Sussex, run by C. Newington, where he remained for almost another two years. After his release, he wrote and published anonymously *A Narrative of the Treatment Experienced by a Gentleman, During a State of Mental Derangement* which appeared in 1838.[36] A second edition of the work, which adds much material and amplifies and particularizes the complaints of his treatment in Dr Fox's institution, appeared with his authorship openly acknowledged in 1840. The two books were edited into a single volume by Gregory Bateson in 1961.[37]

Thereafter, Perceval began an active and life-long campaign against the injustices of the system of incarceration and care of lunatics. Perceval has become something of a hero of the struggle against the cruel and brutal treatment of the mentally ill, though many of his complaints about his own treatment revolve around his indignation that a gentleman such as he felt himself to be should have had to endure the indignity of being banged up with his social inferiors. Perceval's delusions had a strongly religious cast, and consisted very largely of the hearing of voices, which issued a bewildering variety of demands, exhortations and imprecations. But the theme of machinery also runs through his account, though, as with Friedrich Krauß, the machinery of his delusions is perhaps less apparent than in Matthews' because of their predominantly auditory nature.

Perceval seems to have experienced an extreme alienation from his own body, which he represents as acting and acted upon as though it were mechanical. At various times, he feels his body moved automatically, as though by some mechanism, as, for example, when he strikes his neighbour at the tea table: 'My hand struck that blow, but it was involuntary on my part, as if my hand had been moved by a violent wind. A spirit seized my arm with great rapidity, and I struck as if I was a girl'.[38] It is unclear whether this machinery is pneumatically or electrically driven. On another occasion, his right arm 'was suddenly raised, and my hand drawn rapidly across my throat, as if by galvanism'.[39] The most explicit indication of his mechanical sense of his body is his claim that 'My loss of all control over my will, and belief, and imagination, and even of certain muscles, was immediately preceded by three successive crepitations, like that of electrical sparks in the right temple, not on the same spot, but in a line, one after the other, from left to right'.[40]

Perceval seems to have suffered from an extreme alienation from his own thoughts and process of thinking, an alienation that makes him unable to recognize his thoughts and thought processes as his own. But this alienation also seems to give him occasional glimpses of the process or machinery of his thinking, though he is much less precise than Matthews about the nature of the machinery involved. His alienation thus appears to bring him close to consciousness of his own thought:

> I recollect I found myself one day left alone, and at liberty to leave my bed. I got up, and knelt down to pray. I did not pray, but I saw a vision, intended, as I understood, to convey to me the idea of the mechanism of the human mind![41]

There are scattered hints through his testimony that this machinery is at least in part pneumatic. Perceval sees many of his mental disorders in terms of disturbances of the air and breath. 'I have found that whenever my bodily health has been deranged, particularly whenever my stomach has been affected, I have been more than usually troubled by these fancies, particularly if at the same time, through sluggishness or through cold, I have not been breathing through my nostrils, or drawing deep breaths'.[42] One of the strangest of these disturbances is what might be called a *panophonia* – the production of voices out of ordinary sounds, especially the internal sounds of his own body: 'I found that the breathing of my nostrils also, particularly when I was agitated, had been and was clothed with words and sentences'.[43] The sound of air is particularly liable to become, in his expressive phrase 'clothed with articulation'.[44] He describes his fear at the approach of his attendants: 'Their footsteps talked to me as they came up stairs, the breathing of their nostrils over me as they unfastened me, whispered threatenings; a machine I used to hear at work pumping, spoke horrors'.[45] As he begins to recover, he is able increasingly to identify the sources of these sounds: 'I discovered one day, when I thought I was attending to a voice that was speaking to me, that, my mind being suddenly directed to outward objects, – the sound remained but the voice was gone; the sound proceeded from a neighbouring room or from a draft of air through the window or doorway'.[46]

If voices are produced through disturbed respiration, then, Perceval believes, recovery will entail the proper regulation of the machinery of breath: 'I question whether the operations of the conscience and reflection can be conducted but through the medium of the lungs filling the chest at proper intervals, according to the degree of passion of the mind, or of action of the body'.[47] He believes that this disordering can be reversed by conscious control of the breath:

> I believe the healthy state of the mind depends very much upon the regulation of the inspiration and expiration; that the direction "*animum rege*", has a physical as a well as a spiritual sense; that is, in controlling the spirit you must control your respirations. I will instance, in support of this, the stupid appearance of many deaf people, who are usually unable to breathe freely through the nostril, and keep their mouths wide open; a habit very common among idiots. I will instance, again, the stupefying effects of a bad cold.[48]

Perceval suggests that mechanical assistance may help to regulate the breath: 'it is possible, that the effecting of this [breathing] mechanically even may give much relief. I have certainly found it so'.[49]

Elsewhere, Perceval defines madness as the incapacity to distinguish literal from figurative expressions, and the consequent tendency to take metaphors literally. This applies particularly, he thinks, to the voices he heard commanding him to suffocate himself: 'when I was desired to suffocate myself on my pillow, and that all the world were suffocating for me, &c. &c., I conceive, now, that the spirit referred to the suffocation of my feelings – that I was to suffocate my grief, my indignation, or what not, on the pillow of my conscience'.[50] But his emphasis on the literality of breathing, on the physical as well as spiritual sense of '*animus*', indicates that he is still inclined to take the letter for the spirit, especially when it is the letter of the spirit: for Perceval, thought is, more than ever, breath.

Perceval sums up his beliefs in a series of propositions:

> To make my ideas more clear, let me sum up my arguments or propositions thus: That a healthy state of the mind is identical with a certain regulated system of respiration, according to the degree of bodily action; that the exercise of reflection or of conscience, in the control

of the passions or affections of the mind, is concomitant with, or effected by a proper control of the respiration – quiet when the mind is quiet, accompanied with sobs and sighs when otherwise. That the mind and the blood, being intimately connected, the health of the body depends also on this healthy regulation of respiration, promoting a proper circulation and purification of the blood; that, consequently, the effecting respiration by mechanical means, without the control of the muscles by thought, is profitable to the health of the body, and also to that of the mental faculties, although they may not be, at least distinctly, occupied by any ideas; in the same way as, if several printing-presses are worked by machinery, it may be necessary for the perfect state of that machinery, that all the presses should be in motion, though some may have no types under them.[51]

This statement hovers in a characteristic way between conventional wisdom regarding the coordination of breathing and cognition and obsessional literalism. Perceval seems to imagine a kind of iron lung or automatic regulator of the breath to ensure the orderly production of thoughts – here imagined as the output of a printing press. Remarkably, he seems to have no sense of any relation to the import of these thoughts, to which he stands in the same relation as the proprietor of a printing works to the content of what is being printed – and Perceval is drawing attention to the fact that the machine needs to be kept running even when there is no type to be reproduced, or thoughts to occupy the mental faculties. This is the maturing of Perceval's image of his mind as a kind of pulmonary thought-works.

Tausk's influencing machine is the body, and more specifically the genitals, made unrecognizable by the machinations of desire, power and fear. Presumably some part of the therapy and recovery of the psychotic would be the unmasking of the body behind the machine. Perceval's recovery also takes the form of a demystification, or a somatizing of effects that had been thought to be spiritual or supernatural. Voices are explained in terms of the mechanics of auditory hallucination; the exercise of sound judgement is associated with the equipoise of the breath. But Perceval's body is not a deliverance from the obfuscations of the machine, or from the machinery of his delusion, for his is a mechanical body, which is to take the place of the spiritual body of which he believed himself possessed at an early stage of his madness. Perceval sees the understanding of the machinery of his body as the proof of his recovery, when it is in fact the most tenacious aspect of his pathology. The neutral mechanism of his body and breath may take away the authority of his voices and allow him to ignore rather than strive to obey them, but that mechanism is still deeply delusive. In Perceval's case, the turn to the body cannot wholly dissipate the mechanism of his delusions, since his body is a delusive machinery. We should remember that, whereas Matthews' air-loom belongs to the most florid period of his persecutory delusion, these and other theories belong to the period of what Perceval insists is his recovery. Perhaps an externalized influencing machine never materialized in Perceval's thinking because his influencing machine is, in the end, the ideal, regulated mechanism of his own embodied thinking, which he is convinced has delivered him from the disordering of his wits. Recovery for Perceval seems to consist in replacing the idea that he is possessed by spiritual agencies with the idea that he is suffering from a disorder of his mental machinery – but the idea he entertains of this mental machinery is subtly, but tenaciously, delusive. In Matthews' case, the consolidation of the influencing machine was both the fulfilment and the containment of his madness. For Krauß and Perceval, the influencing machine is subdued, not by being destroyed, but by being generalized. Perceval in particular overcomes his influencing machine by becoming it.

Air-machines

Because air is the privileged matter of thought, the machinery of air is a way for thought to think its own workings, encouraged by the development of machines that produced, manipulated and projected across the spaces of the air. The fluid mechanics of air provide a way of embodying the complex machineries of thought, especially in its self-attention. As the fugitive matter of mind, air had always been a way of imaging the irreducibility of mind to body, and even the intractability of mind to itself. But the delusions of 'mechanoiacs' – mechanical paranoiacs – are not so much the signs of a dissolution of a Cartesian subjectivity by telematic media, or of the eruption of 'the unspeakable reaches of his unconscious', as Mike Jay suggests,[52] as the signs of a crisis of hyperconsciousness, a consciousness brought to crisis by the terrifying intensity of its fancied consciousness of itself. So the problem is not one of being alienated from one's thoughts, or dispossessed of them, but rather of coming too close to them. The passion of the machine arises not in the fear of alienation by, but in the voluptuous delight of, identification with it.

Indeed, the apparent terror at possessing agencies may be a systematic defence against the greater terror of having no other way to escape from this excruciating proximity to self. The fantasy of the machine out there is a repudiation of the machine of fantasy in here. It always allows its victim to say 'I am assailed', 'I am worked', 'I am possessed', to preserve the 'I' in the agonized cry 'I am not myself'. It is a defence against the omnipotence of thought, the fearsome identity of thought with itself and the dominion of thought over itself (which means the subjection of thought to itself). It is a problem of reflexivity, the pathology of persecutory self-knowledge. It is for this reason that there is so close a relationship between sickness and therapy in such conditions, both of which take the form of laying bare the machinery of the soul in order to allow the self to exercise mastery over the machine. The influencing machine is a desperate attempt to restore the world, to restore the otherness of the world. Matthews is not trying to save his soul, but to save himself from it. For what does it profit a man if he gain his soul but lose the whole world?

There is a particular difficulty in using an air-machine to restore the determinacy and the outerness of the out-there world. For what is an air-machine? To understand this, we need to distinguish between two conceptions of space. In the more usual conception, bodies are distributed in space, which is both interrupted and contoured by their presence, as a plain or a wilderness is given form and visibility by the trees, hills and lakes that arise in it. Space is articulated, explicated and orientated by the bodies it contains. Existence in space means that these bodies are determinate: they have a particular size, shape and position, and particular relations with other bodies (they are in contact, at a greater or lesser distance, above, below or beside these other bodies). Bodies mark out places in space which is thereby clumped or quantized – 'striated', in Deleuze and Guattari's term. In Mesmer's conception, bodies are not distributed in space: rather space, in the form of the infinitely subtle vital fluid, permeates bodies. Distance, location and orientation are inexistent or meaningless from the perspective of this fluid, which nothing can exclude, obstruct or divide, and which allows every part of space and time to be in contact with every other part. The space of this fluid may be thought of as 'smooth' space in Deleuze and Guattari's terms.

The machine is the paradoxical striation of this smoothness, the impossible container for the universal acid. The fluid that the air-loom brews up and spews out, in what appear to be not merely just blasts or currents but actual threads or filaments, produces a geometry of spaces. But it is precisely the representability of the fluid and the machine, the paradoxical ability of this fluid to be concentrated, channelled, stopped up, bottled, amplified, which also gives Matthews his power against his persecutors, or the possibility of resisting their power over him. Not just the machine, but the act of illustrating it so carefully, may have helped, as Hartmut Kraft suggests, to give a concrete form to Matthews' otherwise formless and deforming agonies of mind and body.[53]

A machine exhausts itself in its operation: its being is its doing. Only a living thing can do more than it does, for instance by meaning to do what it does, or refraining from doing it. To be sure, a machine has potential, indeed it may be thought of as no more than the storing up of such determinate potential. But it has the potential only to perform again what it has already performed in the past. It does not, so to speak, have the potential for possibility which the things we think of as 'living' do.

If, because of its affinity with spirit, air approximates to something like pure possibility, the smooth, unorientated possibility of everything being possible, the air-machine reduces that possibility to particular potentials and protocols – the effects that may be produced at particular times and places and in particular ways. One might say that the mesmerist is already in this sense an air-machine. His techniques and routines, and the apparatuses like the *baquet* which allow them to be exercised automatically, are the ways in which the vital flow can be regulated, by the introduction of differential stresses and tensions into the otherwise absolutely smooth continuum of all-pervasive, resistless fluid. The machine, the machinery, the operator are all kinds of battery – which, in the form of the Leyden jar, had been invented only a few decades previously.

The air-machine turns quality into quantity, force into substance. It makes the ineffable fluid finite and manipulable. This is because, in its essence, the machine has no secrets, no hidden interiority, no nonfunctional residuum, no quiddity that is not accounted for in the details of its operation. To be sure, there are complex, inefficient and exhibitionist machines, machines that seem to consist of show; but in such devices, these are inessential elements that do not belong to the machine itself. The essence of the machine is that it is has no essence separate from its action, and is thus finitisable, totalisable.

But there is also a kind of wild or infinite machine, which exists as a generalized and proliferating machinery, a machine without a limit, that propagates rather than exhausting itself in its operation. This kind of machinery does not coincide with what it does: it includes and exceeds what it does. Machines are demonstrative in their nature: they can be opened out and revealed. But the air-loom is a black box: it cannot be seen.

The infinite machine is at once a desublimation and a remystification of thought. Conceived of as 'mere' machinery, the mind is reduced to material, mechanical operations. But the growing power invested in the idea of the machine, which increasingly could be thought of not just as supplementing human actions, but as displacing them, gave it a new, quasi-magical autonomy. The more that thought became automatic, the more autonomous that automaticity could become. The air is increasingly subject to mechanics – but, in Matthews' imagination, the machine had already begun melting into air.

So there must be two kinds of air-machine. One is bulky, odorous, corporeal, kinetic. This is set against another, which is volatile, edgeless, self-generating. It is in fact a bad infinity, part of the infinite machine, or dissolution of the world into thought. This is not a defence of the infinitude of the air against its finitizing, but a defence of the concrete against the abstract air, the air of breath against the air of universal thought. This explains the desire to materialize, or mechanize, the self, in archaic machines that are defences against the infinite machinery of thought. Matthews, Krauss, Perceval and Schreber do not want to be integrated with their thoughts, they want to be smaller than their thought. The thought 'nothing is external to me' is made over into the thought 'everything is external to me'. The fear of becoming everything is disguised and deflected by the fear of becoming nothing. The archaic machine is a body in space, which explains the importance of the white space in Matthews' drawing. As the influencing machine is elaborated, it will expand to occupy more and more of the available space – pictures by other sufferers from a later period show a much more saturated space, full of wires and interconnections. Matthews' machine is incomplete, and he has directed that some parts of the illustration be left sketchy. This may suggest both that the machine has not yet been fully revealed to him and also that the machine itself is dissolving into the air that it itself weaves – as though the machine were indeed beginning to become everything, to take to the air. The yawning spaces of Matthews' illustration, and the incompleteness to which they testify, are a way of keeping himself apart from himself, of forestalling his identification with and as 'Bill the King', the leader of the gang who is nowhere to be seen in the illustration.

The mechanization of air doubles and assists the corporealization of thought. But the resulting body of thought is an air-body, made material in the image of what was beginning to become the reference-state of matter – that is, the volatile, the vaporous, the airy. In the process, the machine, too, changed its character. It became generalized, indefinite, ubiquitous, self-generating. These men came early upon a machinery that had started taking to the air.

Notes

1. Victor Tausk, 'On the Origin of the "Influencing Machine" in Schizophrenia', in Paul Roazen (ed.), Dorian Feigenbaum (trans.), *Sexuality, War and Schizophrenia: Collected Psychoanalytic Papers* (New Brunswick: Transaction Publishers, 1991), 187.
2. Tausk, 'On the Origin of the "Influencing Machine" in Schizophrenia', 208.
3. Tausk, 'On the Origin of the "Influencing Machine" in Schizophrenia', 210.
4. Mike Jay, *The Air-Loom Gang; The Strange and True Story of James Tilly Matthews and His Visionary Madness* (London and New York: Bantam Books, 2004).
5. John Haslam, *Illustrations of Madness* (1810; repr., London and New York: Routledge, 1988), 19.
6. Roy Porter, Introduction to *Illustrations of Madness*, by John Haslam (London and New York: Routledge, 1988), lxiii.
7. Haslam, *Illustrations of Madness*, 45.
8. Haslam, *Illustrations of Madness*, 45.
9. Haslam, *Illustrations of Madness*, 48.
10. Haslam, *Illustrations of Madness*, 43.
11. Haslam, *Illustrations of Madness*, 45, 42–43.
12. Haslam, *Illustrations of Madness*, 30–1.
13. Haslam, *Illustrations of Madness*, 32.
14. Haslam, *Illustrations of Madness*, 35.
15. Haslam, *Illustrations of Madness*, 37–38.
16. Haslam, *Illustrations of Madness*, 38.
17. Haslam, *Illustrations of Madness*, 28–9.
18. John Haslam, *Observations on Madness and Melancholy: Including Practical Remarks on Those Diseases; Together With Cases: And An Account of the Morbid Appearances on Dissection*, 2nd edn (London: for J. Callow, 1809), 9.
19. Haslam, *Observations on Madness and Melancholy*, 9.
20. Haslam, *Observations on Madness and Melancholy*, 34.
21. Haslam, *Observations on Madness and Melancholy*, 269.
22. Friedrich Krauß, *Nothschrei eines Magnetisch-Vergifteten (1852) und Nothgedrungene Fortsetzung meines Nothschrei (1867): Selbstschilderungen eines Geisteskranken*, ed. H. Ahlenstiel and J.E. Meyer (Göttingen: Bayer-Leverkusen, 1967). Further selections appear in *Grenzgänge zwischen Wahn und Wissen: zur Koevolution von Experiment und Paranoia, 1850–1910*, ed. Torsten Hahn, Jutta Person and Nicolas Pethes (Frankfurt and New York: Campus, 2002), 35–57.
23. Krauß, *Nothschrei eines Magnetisch-Vergifteten*, 19.
24. Krauß, *Nothschrei eines Magnetisch-Vergifteten*, 14.
25. Friedrich Krauß, quoted in Stefan Rieger, 'Psychopaths Electrified: Die Wahnwege des Wissens im *Notschrei eines Magnetisch-Vergifteten*', in Torsten Hahn, Jutta Person and Nicolas Pethes (eds), *Grenzgänge zwischen Wahn und Wissen: zur Koevolution von Experiment und Paranoia, 1850–1910* (Frankfurt and New York: Campus, 2002), 163.

26. Krauß, *Grenzgänge zwischen Wahn und Wissen*, 41.
27. Krauß, *Nothschrei eines Magnetisch-Vergifteten*, 25.
28. Krauß, *Nothschrei eines Magnetisch-Vergifteten*, 23.
29. Krauß, *Nothschrei eines Magnetisch-Vergifteten*, 14.
30. Krauß, *Grenzgänge zwischen Wahn und Wissen*, 50.
31. Friedrich Krauß, quoted in Bernhard Siegert, 'Gehörgänge ins Jenseits: Der telephonistische Entzug des Ohres', in Torsten Hahn, Jutta Person and Nicolas Pethes (eds), *Grenzgänge zwischen Wahn und Wissen: zur Koevolution von Experiment und Paranoia, 1850–1910* (Frankfurt and New York: Campus, 2002), 181.
32. Friedrich Krauß, quoted in Christine Wunnicke, '"Auserwählt zum Aufbruch": Der bürgerliche Wahsninn des Friedrich Krauß', in Torsten Hahn, Jutta Person and Nicolas Pethes (eds), *Grenzgänge zwischen Wahn und Wissen: zur Koevolution von Experiment und Paranoia, 1850–1910* (Frankfurt and New York: Campus, 2002), 120.
33. Krauß, *Grenzgänge zwischen Wahn und Wissen*, 46–48. See D.G. Kieser, *System des Tellurismus oder thierischen Magnetismus: Ein Handbuch für Naturforscher und Aerzte*, 2nd edn (Leipzig: F.L. Hervig, 1826).
34. Krauß, *Grenzgänge zwischen Wahn und Wissen*, 50.
35. Krauß, *Grenzgänge zwischen Wahn und Wissen*, 37.
36. John Perceval, *A Narrative of the Treatment Experienced by a Gentleman, During a State of Mental Derangement: Designed to Explain the Causes and the Nature of Insanity, and to Expose the Injudicious Conduct Pursued Towards Many Sufferers Under That Calamity* (London: Effingham Wilson, 1838).
37. John Perceval, *Perceval's Narrative: A Patient's Account of His Psychosis 1830–1832*, ed. Gregory Bateson (Stanford: Stanford University Press, 1961).
38. Perceval, *Perceval's Narrative*, 113–14.
39. Perceval, *Perceval's Narrative*, 118.
40. Perceval, *Perceval's Narrative*, 284n.
41. Perceval, *Perceval's Narrative*, 54.
42. Perceval, *Perceval's Narrative*, 298.
43. Perceval, *Perceval's Narrative*, 295.
44. Perceval, *Perceval's Narrative*, 265.
45. Perceval, *Perceval's Narrative*, 93.
46. Perceval, *Perceval's Narrative*, 294.
47. Perceval, *Perceval's Narrative*, 272.
48. Perceval, *Perceval's Narrative*, 271–72.
49. Perceval, *Perceval's Narrative*, 272–73.
50. Perceval, *Perceval's Narrative*, 271.
51. Perceval, *Perceval's Narrative*, 273.
52. Jay, *The Air-Loom Gang*, 226.
53. Hartmut Kraft, *Grenzgänge zwischen Kunst und Psychiatrie* (Cologne: Dumont Verlag, 1986), 55–56.

3
Maternity, Madness and Mechanization: The Ghastly Automaton in James Hogg's *The Three Perils of Woman*

Katherine Inglis

The meanings of the automaton

Of all James Hogg's works, *The Three Perils of Woman: Love, Leasing and Jealousy* (1823) has the strongest claim to being the most unpopular.[1] At the time of publication, it was a critical and commercial failure, condemned as vulgar, coarse, indecent and utterly unsuitable for female readers.[2] A brief survey of some of the events in the novel shows why in 1823 this was inevitable: this is a novel in which a sentimental heroine becomes a living corpse, that rewards a 'fallen' woman with a happy marriage, that represents, without sentiment, the brutalization of the rural population by war, and ends with an insane mother singing to her dead newborn. Much of Hogg's work was reissued after his death in a sanitized, Anglicized form (even *The Private Memoirs and Confessions of a Justified Sinner* was reissued for Victorian readers) but *Perils of Woman* proved irrecoverable. Such a novel cannot be 'bowdlerized and domesticated'.[3] If an editor were to excise all the elements with potential to offend, there would be no novel: specifically, there would be no mothers.

Events in *Perils of Woman* suggest that maternity is the most profound peril for women. Ostensibly, love, leasing (lying) and jealousy provide the novel's structural and thematic organisation. The first 'peril', 'Love', a version of the 'national tale' genre, is set in modern Edinburgh and its environs; 'Leasing' tells the story of Sally, a serving-maid prone to lying, in the period leading up to Culloden; 'Jealousy' shows the aftermath of the massacre and Sally's descent into madness. Yet this tripartite organization is undone by the novel's irregularity and excess: the named perils creep into other tales, the second and third tales are effectively a single novella, and the first peril dispenses with its main

61

plot and genteel characters, concluding instead with an extended epis-
tolary comic digression. The perils of maternity, however, draw the
'perils' and disparate plots together. One could be forgiven for thinking
that maternity was a pathological condition or a crime: the two principal heroines fall into a state of incoherence and abjection during their
pregnancies; mothers are cast into asylums, imprisoned, threatened
with forcible restraint; neonates are stolen and die. Anxiety toward
the maternal body, and recurring images of maternal trauma, connect
the three perils, yet the maternal body, being a locus of trauma, must
also be repressed. Maternity is at once the novel's principal concern
and its shameful secret. The generation of children happens out of
sight, at the margins of society and geography. The heroine of 'Love',
Agatha (Gatty) Bell, gives birth in a private asylum during a three-year
period of unconsciousness. Mrs Johnson, Gatty's nurse, thinks that
she gave birth to a still-born child, when in fact her living child was
stolen from her. Katie, the heroine of the comic sub-plot, is threatened
with a straightjacket by her seducer, who intends to force her to give
up her son. Sally gives birth in the wilderness as a widow. Newborns
are brought into the world not in triumph, through the *labour* of the
mother, but *delivered* mysteriously *from* silent bodies with mechanical
efficiency. Or rather, from bodies that are incoherent. They produce
language, but a form of language that proves incomprehensible to their
auditors, thus frustrating the drive to invest the figure of the mother
with meaning. The maternal body is an uncanny form in Hogg's novel:
that which in the ideology of the national tale or the historical novel
is the repository of national meaning, agency and continuity, becomes
instead an emblem and agent of the disruption of history.[4] The canny
becomes the uncanny. At the root of this shift in meaning is the representation of Gatty as a 'ghastly automaton' in the first peril.[5] The
meanings of 'automaton' are – like the meanings of 'uncanny' – by
turns fluid and antithetical. These unstable meanings shape the formation, function, and spectacular dysfunction of Hogg's mechanical
mothers.

The term 'automaton' expresses two contradictory meanings:
self-moving and directed. Though Freud paid the automaton scant
attention, finding its uncanny potential less than convincing, the
etymological evolution of 'automaton' parallels that of 'uncanny'
from 'canny'. *Heimlich*, like 'automaton', expresses two distinct sets
of ideas: 'what is familiar and agreeable' (literally, homely) and 'what
is concealed and kept out of sight', an idea that develops from the

first, as in 'withdrawn from the eyes of strangers' or 'withdrawn from knowledge'.[6] The progression of *heimlich* from 'homely' to 'concealed' culminates in the establishment of a sense that is synonymous with *unheimlich*, something that is *un*familiar and *dis*agreeable, secret or occult. A similar ambiguity exists in the Scots word *canny*, which can mean variously 'safe', 'lucky', 'sagacious', 'of good omen' and – developing from this last usage – 'one who deals with the supernatural'.[7] So, paradoxically, a 'canny wife', a wise woman, is *nae cannie*. Thus the meaning of the word *heimlich*, Freud demonstrated, 'develops in the direction of ambivalence, until it finally coincides with its opposite, *unheimlich*', so that *unheimlich* is a 'subspecies' of *heimlich*.[8] Following a similar pattern, 'automaton', which originally meant 'self-moving', has over time 'accrued the meaning of an apparently self-directing process which in reality has its motion determined'.[9] Nineteenth-century definitions of 'automaton' pair these antithetical meanings as awkwardly conjoined twins. The *Edinburgh Encyclopaedia*, which was edited by Hogg's acquaintance David Brewster, defined 'automaton' as 'a self-moving machine, or machine so constructed, that, by means of internal springs and weights, it may move a considerable time as if endowed with life'.[10] So the automaton is at once a true self-mover and a mimic of autonomous motion, a mechanical simile that moves *as if* endowed with life. Thesis slips into antithesis, then comes to be constituted *as* its antithesis. The automaton is defined by its lack of autonomy, thus to be automatous (like an automaton) is to be reduced to a mechanistic semblance of agency. Gatty's affinity with the paradoxical self-mover reveals the subjugation of her consciousness to a terrifying tyrannical body. That this illustration is made through metaphor rather than simile (the trope most commonly associated with automata) demonstrates the extremity of her degradation: she becomes an automaton indeed, not *like* an automaton. Her fate haunts all mothers in *Perils of Woman*. Maternity is, in a very real sense, ghastly; the conversion of mothers to machines barely metaphorical. However, ghastliness is succeeded by an alternate way of reading the automaton. The *Edinburgh Encyclopaedia* defined the 'androide' as a superior automaton:

> A machine resembling the human figure, and so contrived as to imitate certain motions or actions of the living man. It is considered as the most perfect or difficult of the *automata* or self-moving engines; because the motions of the human body are more complicated than those of any other living creature.[11]

The most perfect of machines overlaps with the most complex of living forms. In the closing pages of the first peril the ghastly automaton becomes an ideal humanoid form in an attempt to render Gatty's mechanization palatable, a recuperative interpretation that has been read as if it were transparent, but is treated in the most recent criticism of the novel with suspicion. This narrator's voice is perhaps no more trustworthy than those of the discordant narrators of *The Private Memoirs and Confessions of a Justified Sinner*, which was published the year after *Perils of Woman*.

Hogg's automaton is both a fatal possibility and an ideal, two radically different meanings that can be aligned with two disciplines that imagined the human body as an automaton, respectively galvanism and midwifery. The ideal automaton appears in the awkward coda to Gatty's 'peril', wherein her husband and the colluding narrator attempt to rehabilitate the ghastly automaton, recasting her *de*generation as a mysterious process of generation, shifting the mechanical mother from the realm of the monstrous to the sublime. Her abject, automatous body is re-imagined as a superbly functional body that performs its work of reproduction more efficiently than it could have done with the distraction of consciousness. It is an extreme, gothic reworking of the representation of the maternal body as a machine in eighteenth-century midwifery literature and medical illustration. Led by William Smellie, whose midwifery courses reduced childbirth to mechanical principles, British midwifery literature imagined the female body as a complex machine (like the android, the most perfect and difficult of all automata), and used automata to represent generation and parturition. The automatous mother became an ideal in representation and practice. Hogg's fusion of midwifery rhetoric and galvanism drags the automaton down to the level of a dissected electrified corpse, casting doubt on the optimistic interpretation provided by Gatty's husband and the narrator. The galvanized corpse threatens to recur (or reanimate) throughout the novel. The mechanization and objectification of the mother is a terrifying possibility in the novel's comic plot that is tragically fulfilled in the course of the second and third perils. Extraordinary acts of violence are done to the bodies of women and children in imagination and in fact: Gatty is likened to a razor blade that should be ground down; a gravedigger slices the lobes off the ears of living children in order to obtain payment for burials he has not performed; communities are laid waste in the aftermath of Culloden. Razors, slicing, butchery: a lexicon of anatomical dissection reverberates through the novel, bringing the idealization of the automaton into question.

'As If Endowed With Life': the galvanized corpse

Two fantasies of reduction predict the coming of the automaton. Gatty foretells her death, promising she will be 'lying on that bed a lifeless corse' by the next Sunday.[12] She refuses to accept 'apothecary's drugs, these great resorts of the faithless and the coward'.[13] Daniel, her father, responds to her rejection of medicine with a prophecy of his own:

> Your spirit has often brought me in mind of a razor that's ower thin ground, an' ower keen set, whilk, instead of being usefu' an' serviceable, thraws in the edge, or is shattered away til a saw, an' maun either be thrown aside as useless, or ground up anew. Now, my dear bairn, an this thin an' sensitive edge war ground off ye awee on the rough hard whinstone of affliction, I think ye will live to be a blessing to a' concerned wi' ye.[14]

These two fantasies – of transformation into a corpse and the grinding of metal – are welded together in Gatty's 'death' and reanimation. She becomes a curiously metallic corpse (if not a '*lifeless* corse'), and thereafter a useful and serviceable version of herself. Her sharp edges ground off, Gatty becomes Agatha, 'a blessing to the human race' as Daniel predicted – but at a cost.[15] Gatty's plea was for autonomy, to decide the manner of her death by refusing medical treatment, but Daniel's rhetoric of reduction and utility discounts agency entirely. His narrative of destruction and reconstruction is imposed retroactively and governs the way in which Gatty's ordeal will be understood. Her physical reduction evolves into the reduction and rewriting of her identity.

The grinding of the razor begins with the secret introduction of medication to the death-chamber. Though Gatty had denied the apothecaries, she remains accessible to one surgeon, her husband. Fearing that her apprehension of death will prove fatal, M'Ion administers 'small portions of a cordial elixir . . . sweetened and diluted with wine and water' (laudanum) so that she will sleep through the expected moment of death.[16] His stratagem fails. She sinks almost imperceptibly into a deep sleep, and from that to what appears to be death. There are no signs of life, and she is dressed for burial. However, her husband and father (against the wishes of the women laying out the body) force their way into the death chamber and say prayers that approach blasphemy, demanding her resurrection. It is left unexplained whether what ensues is a punishment for their presumption or a consequence of the laudanum.[17] Both acts – the secret administration of laudanum and the intrusion – compromise female agency. It is fitting that the female body should respond by asserting itself as sheer physical force. At this point, readers might have expected one of two possible issues: an edifying sentimental death or a fairytale resuscitation in which Gatty is woken by the kiss of her Highland Prince Charming; however, the narrative swerves from these established tracks into a little-trodden path – the outrageous mechanization of the sentimental heroine. She revives, but as an uncanny version of herself, a body denuded of mind: incoherent, aggressive and infused with kinetic and electrical energy.

Recent critical and biographical studies of Hogg, and the Stirling/South Carolina editions of his collected works, have drawn attention to his informed, imaginative response to the Scottish Enlightenment, demonstrating conclusively that the strangely persistent caricature of Hogg as a superstitious, inspired shepherd-poet, who by some mystery was able to write a single, atypical masterpiece, is a fallacy. In Ian Duncan's study of the novel in Romantic Edinburgh, *Scott's Shadow* (2007), Hogg's fiction is acclaimed as that which is most aware of and responsive to scientific and philosophical progress. Hogg was 'imaginatively more attuned to the intellectual currents of advanced modernity, including radical materialism, than any contemporary Scots author'.[18] This imaginative refashioning of radical materialism is evident in Hogg's unsettling representation of bodies as uncanny *things*. Hogg's uncanny is constituted not of the ethereal sublimity that might be expected of the self-proclaimed 'king o' the mountain an' fairy school' of poetry, but of 'a weird insistence of physical bodies' in the absence of mind.[19] In *Perils of Woman*, Hogg re-imagines galvanism, playing with its language and anxieties to create a nightmare version of the galvanic body. Reanimation is configured as galvanic theatre: shortly after M'Ion has recognized that a 'spark of life remains', M'Ion and Daniel each place a hand on Gatty's chest – their hands act like galvanic contact points, conducting their blasphemous prayers as animating current.[20] His sentimental heroine becomes an 'upright corpse', her continued existence expressing the 'the horror of a mechanistic pseudolife, an animation unendowed with reason or sensibility'.[21] Stripped of agency, reduced to an 'object', the heroine persists in mechanical form.[22]

The danger of galvanism, as its practitioners realized, was that by showing that the body retained a degree of function in the absence of consciousness or spirit, it suggested a resemblance between the human body and a machine. Giovanni Aldini, nephew of Luigi Galvani, protested against this implied kinship:

> I should think it a prostitution of galvanism, if it were only employed, to cause sudden gestures, and to convulse the remains of human

bodies; as a mechanic deceives the common people by moving an automaton by the aid of springs and other contrivances.[23]

To align the human with the automaton, that mechanical dissembler, is to debase galvanism; but the deeper, unspoken fear is that galvanic gesture might not be mere deception, but actually equivalent to genuine animation. *Perils of Woman* imagines this extreme outcome of galvanic experimentation. 'Lacking spirit, reason, or voice, the reanimated corpse is a body reduced to a thing, a horrifying nexus of matter and force'.[24]

There are marked similarities between Hogg's reanimated corpse and Andrew Ure's account of his experiments on the body of a man hanged for murder in Glasgow in 1818.[25] In both narratives, a sentient, named individual becomes an anonymous, mechanistic and vigorously theatrical body. Gatty loses her identity, becoming a nameless 'it', just as in Ure's account the murderer Clydesdale became 'the body' or 'the subject'. Ownership of the body and its narrative passes from the subject to the anatomist/medical practitioner. Ure's first, unabridged account, delivered to the Glasgow Literary Society in 1818 and subsequently published in 1819, is positively theatrical. His style is unstable, switching abruptly from a clinical description of his exposure of the nerves to be galvanized, to a dramatic rendering of the body's gruesome convulsions and the audience's horror.[26] When Ure applied current to the heel and the supra-orbital nerve in the forehead, the 'most extraordinary grimaces were exhibited':

> Every muscle in his countenance was simultaneously thrown into fearful action; rage, horror, despair, anguish, and ghastly smiles, united their hideous expression in the murderer's face, surpassing far the wildest representations of a Fuseli or a Kean. At this point several of the spectators were forced to leave the apartment from terror or sickness, and one gentleman fainted.[27]

The corpse has become Ure's artistic creation, a spectacle to be gazed at like a sensational painting or a theatrical performance. The mechanistic implications of Ure's staging become clear when he describes setting the pulmonary organs 'a-playing': the body is effectively a musical instrument, control of which has passed from Clydesdale to Ure.[28] Clydesdale's body may be perpetually on the verge of coming to life, but its incipient animation is overwhelmed by its reduction to an instrument. Clydesdale becomes an organic automaton manipulated for the edification and entertainment of Ure's audience.

The staging and progress of Gatty's reanimation replicates Ure's account: a supine body is displayed before an audience, performs a sequence of grotesque contortions, comes into violent contact with a spectator, and terrifies the audience:

> Behold the corpse sat up in the bed in one moment! The body sprung up with a power resembling that produced by electricity. It did not rise up like one wakening out of a sleep, but with a jerk so violent that it struck the old man on the cheek, almost stupefying him; and there sat the corpse, dressed as it was in its dead-clothes, a most appalling sight as man ever beheld. The whole frame appeared to be convulsed, and as it were struggling to get free of its bandages. It continued, moreover, a sort of hobbling motion, as if it moved on springs. The women shrieked and hid their faces, and both the men retreated a few steps, and stood like fixed statues, gazing in terror.[29]

Electrified, convulsive, moving like a puppet on springs – Hogg's aggressive corpse is mimicking Ure's. Ure recorded that galvanism of the spinal marrow and sciatic nerve produced 'convulsive movements, resembling a violent shuddering from cold', while switching from the sciatic nerve to the heel caused the corpse's leg to be 'thrown out with such violence, as nearly to overturn one of the assistants, who in vain attempted to prevent its extension'.[30] Wrestling matches between the violent corpse and an assistant were frequent in galvanic experiments.[31] Hogg includes this feature of galvanic theatre in the form of a wrestling match between Gatty and her mother, who attempts unsuccessfully to restrain the body, which 'felt as if it were endowed with unnatural force, for it resisted her pressure, and rebounded upwards'.[32] Gatty's affinity with mechanism, and with Clydesdale, is confirmed when Hogg describes her as 'this ghastly automaton':[33] the galvanized corpse's smile was also 'ghastly'.

It is significant that Gatty is identified as a 'ghastly automaton' in the moment her mother accepts the terrible object is 'the body of my child, although it appears that the soul is wanting'.[34] Gatty's condition is characterized by loss, lack, absence – by that which is wanting. The automaton is the antithesis of what it impersonates. Gatty is defined by what she is not, as an '*in*comprehensible being', 'a face *without* the least gleam of mind', 'the poor *remains*', and 'the *shattered* and *degraded* frame of his poor wife'.[35] Even her failed attempts at speech are described in terms of negation: she produces 'a loud and *un*intelligible noise' and

'an articulation that sounded like *"No-no-no!"*'.[36] This embodied negation of all that is human is moved secretly to a private asylum, where
she 'became as a thing altogether forgotten'.[37]

Superficially, Gatty's tale ends with a satisfactory resolution. After
three years, her mind returns, and she leaves the asylum in a condition
that is claimed to be superior to her original state. Her automatism is
now revalued by her family (and confirmed by the narrator) as a transformative process that fitted her for her role as a wife and mother. The
ghastly automaton produced a sedate matron from the nervous young
bride, curing her of the religious enthusiasm that had interfered with
her relationship with her husband, fattening her into a beauty, and most
impressively, turning her into an insensate incubator. It is important to
note at this point that this perspective is not necessarily Hogg's, whose
fiction, particularly *Perils of Woman*, has suffered from the attention of
reviewers who were either deaf to irony or could not believe that the
Ettrick Shepherd was capable of irony. The sudden irruption of Gatty's
violently corporeal body into a parody of a sentimental death-scene *is* a
subversive, ironic act, an audacious riposte to the literary representation
of female death as 'aesthetically pleasing' and 'pictorial'.[38] This ghastly,
ironic automaton will prove an insurmountable challenge to the two
optimistic interpretations that form the coda to its tale.

'The Most Perfect or Difficult of the *Automata*': The Ghastly Incubator

If the family's revaluation of the automaton is accepted, and the narrator's conclusion read as if it is transparent, then Gatty's reformation
appears to be a 'happy ending', the 'triumph of good nature and common sense over excessive and overblown concepts of delicacy and hon-
our'.[39] The razor has been ground up anew, and thank goodness: 'If ever
there was a woman redeemed from the gates of death to be a blessing to
the human race, it has been Agatha Bell'.[40] Yet given the brutal, macabre
method by which Gatty's 'improvement' has been achieved, this per-
functory eulogy seems an inadequate conclusion. As Douglas Mack has
noted, the narrator concludes uncertainly with a question 'rather than
a ringingly confident assertion' of Gatty's perfection: 'Who can doubt
that the Almighty will continue to bless such a benign creature to the
end, and her progeny after her?'.[41] It is always wise to treat Hogg's narrators and their assertions with suspicion. 'Perhaps the conventional
happy ending of Gatty's story is not as straightforwardly happy as it
seems, if one is willing to look beyond the agreeable and respectable

surface of events'.[42] Beneath the agreeable surface of the renovated blessing lies the ghastly body of the automatous mother.

The chronology of Gatty's illness and recovery suggests that she descended into automatism immediately after falling pregnant. Gatty's mind returns after three years, at which point her child is 'two years and three months' old.[43] It was the automaton that gave birth, or rather, was delivered of a child:

> In due time this helpless and forlorn object was safely delivered of a son, without manifesting the slightest ray of conscious existence, or of even experiencing, as far as could be judged, the same throes of nature to which conscious beings are subjected.[44]

If the maternal body is a 'teleological figure' in the 'national tale', achieving continuity and resolution through marriage and reproduction, then Hogg's uncanny maternal body 'unmakes' the genre she represents.[45] In a conventional national tale, the birth of Colin M'Ion would bring resolution by uniting the Lowland Bells with M'Ion's Highland estate, but the mechanical method of (re)production destabilizes this resolution. At this time, birth without pain was associated with 'less civilized' women, who were believed to be either totally insensible to labour pain, or less sensible than hypersensitive 'civilized' women.[46] Gatty's unconscious delivery exposes a natural woman concealed within the rarefied heroine, and reveals that natural woman to be mere mechanism. It would seem that an automaton can be as effective a teleological figure as a national heroine. The 'insensate, mechanical energy' of the body is sufficient 'for the gestation of a child. The body without spirit is a maternal one, the deficiency far from impairing (quite the contrary) its occult powers of procreation'.[47] Hogg's irony is brutal. Denuded of perception, sensation, mind and identity, the essential female body attains its full potential. Pared down to mechanism, Gatty becomes a perfect incubator – passive, undemanding, unconscious. Civilization, sensitivity, intellect and subjectivity prove irrelevant to female identity in Hogg's gothic reworking of the national tale. His principal target is a literary form – the national tale – but Hogg's automatous heroine also recalls the fantasies of mechanical reproduction in eighteenth-century midwifery literature.

In 1821, Hogg and his wife Margaret moved to Edinburgh for the last part of her first pregnancy. She was attended by Dr John Thatcher, man-midwife, lecturer in midwifery and director of a lying-in institution.[48] Hogg was thus well placed to learn about recent developments in midwifery. In the 1820s, the University of Edinburgh made it mandatory for all medical students to attend classes taught by the Professor of Midwifery and the Diseases of Women and Children, threatening the livelihood of independent lecturers like Thatcher. Given that students were already studying midwifery with independent tutors, he argued there was no 'imperious necessity to hurry through this improvement'.[49] Beyond Edinburgh, the nascent profession of obstetrics was moving away from the conservative practices that had dominated from the 1770s. The shift to conservatism in the 1770s was driven partly by man-midwives' attendance at a greater number of onset calls and normal deliveries (previously, they were most likely to be summoned in emergencies to perform radical intervention), but equally by the expectations of an aristocratic clientele unlikely to tolerate intrusive examinations.[50] The conservative school emphasized the mysterious activity of nature over the labour of the man-midwife and instrumental intervention.[51] The disastrous management of the pregnancy of George IV's daughter in 1817 prompted British midwives to rethink their suspicion of intervention and the use of instruments.[52] David Daniel Davis, appointed royal accoucheur in 1819, published *Elements of Operative Midwifery* in 1825, shifting the consensus back to 'balanced management and selective instrumental delivery, which would have gladdened the heart of William Smellie', the mid-eighteenth-century man-midwife who had worked to reduce childbirth to mechanical principles.[53] A revised student's edition of Smellie's *Set of Anatomical Tables* was published in 1823. If Hogg's fictional mechanical maternal body was not responding to the contemporary resurgence of mechanical midwifery, then it was a remarkably serendipitous creation.

Midwifery courses were taught through a variety of methods besides observation of live birth: lecturers used preserved specimens, wax models and mechanical representations of the anatomy of the gravid uterus and female pelvis. These devices were known as automata, mock-women, machines and, towards the end of the nineteenth century, phantoms. Those created by William Smellie were perhaps the most complex ever constructed. One of his students remarked that they were 'so natural' there was little difference between the machines and real women.[54] No images remain of Smellie's automata, but descriptions survive in course prospectuses, auction records, and students' lecture notes. An auctioneer's catalogue published in 1770 lists among Smellie's personal effects four machines (one disassembled), four artificial uteri (two made of glass, one opening with a hinge, one made of leather), and nine artificial foetuses (one 'pretty much used').[55] Each machine had a distinct pedagogical function. One demonstrated natural labours and those made difficult by the circumstances of the child; another more elaborate machine showed the difficulties caused by a narrow pelvis, together with the complete anatomy of the pelvic cavity; a third represented 'all the different Bowels of the Abdomen' and the dilation of the uterus.[56] The prospectus for Smellie's course promised that all variety of natural, difficult, and preternatural labours would be 'perform'd on different Machines made in Imitation of real Women and Children'.[57] Students learned how to perform examinations using machines, and in the fifth lecture 'Each Pupil on a Machine delivers a Child coming in the natural Way, inclosed in the *Uterus*, and surrounded with its Membranes and Waters'.[58] Smellie's lectures were designed around the machines, with 'almost every observation' referring 'to the workings of those machines'.[59] An extract from an anonymous student's notes on the fifth lecture shows the machines' centrality to Smellie's teaching method:

> I have several artificial women every part of which is made to resemble as exactly as possible what is observable in a natural subject. Having laid one of these women on her back with her head and shoulders a little elevated by pillows, her nates brought to the edge of the couch, & her legs & thighs raised towards the abdomen, I then shew that the os Tincœ is open'd at the Time of Birth and by some previous Pains of the mother, about the Bigness of half a crown, thro this aperture can plainly be felt the membranes . . . By each successive Throw of the mother the os Tincœ is more open'd the membranes waters and child's head are push'd further out . . . When the os Tincœ is sufficiently open'd to allow the Child's Head to come out, then we must open the membranes, upon which part of the waters gush out, and along with them part of the Head, we must immediately lay hold of it, & at the next return of the throws gently draw it forward.[60]

Smellie's machines were still used after his death. A Dublin lecturer, Edward Foster, advertised in 1774 that he had procured a machine 'at considerable Expence, and with much Trouble': the 'true Doctrines of Midwifery' were, Foster claimed in the spirit of Smellie, 'at first only intelligible by an Apparatus'.[61]

Smellie's machines polarize modern scholarship on the man-midwifery debates. There is little common ground between histories of the emergence of obstetrics, in which Smellie is a heroic figure, and histories of the suppression of traditional, female-dominated midwifery,

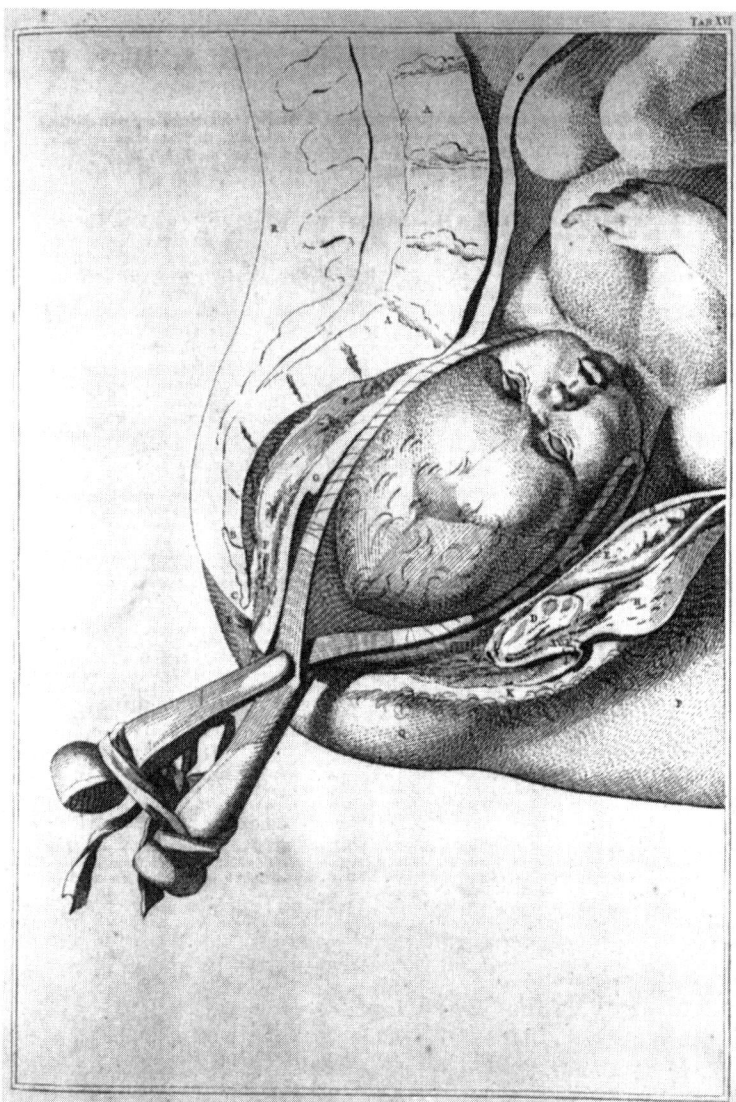

Figure 4: 'Forceps Delivery' from William Smellie's *Tables* (Tab XVI, 1754)

which note that the rise of the man-midwife proceeds alongside the disappearance of the mother as a subject from midwifery literature. Earlier midwifery literature described 'the subjective experience of pregnancy' and showed the mother's entire body, but the only complete subject in

Smellie's representations is the child.[62] Critical appraisals of Smellie's project take his machines to epitomize a certain kind of clinical detachment from the experience of the mother that reduces her to an organic incubator.[63] Arguing that Smellie's mechanical mother acclimated the man-midwife to the idea of a clockwork delivery, Bonnie Blackwell concludes that the 'technological innovations of eighteenth-century obstetric education actually set man-midwives at odds with all but the most inert, the most passive, and the most mechanical of mothers'.[64] In a similar vein, Andrea Henderson argues that the engravings in Smellie's *Set of Anatomical Tables*, which illustrates the progress of labour, represent the mother not as an active participant in labour, 'but *only* as a machine – and an oddly inactive and poorly constructed one at that'.[65] In contrast to William Hunter's realistic images of dissection in *Anatomy of the Gravid Uterus*, Smellie's *Tables* depict bone and clearly differentiated tissue, as in Figure 4, which illustrates forceps delivery. The *Tables* are the visual equivalent of his aspiration to do 'something towards reducing [Midwifery], into a more simple and mechanical method than has hitherto been done'.[66] A representational style is easy to understand, but has, as William Hunter remarked, 'the hardness of a geometrical diagram'.[67] This mechanical style, Henderson argues, effaces the mother's subjectivity. Smellie's 'reader is unlikely to ascribe agency to the mother precisely because she never appears in them as a whole being': the *Tables* show the mother as a series of fragments, and are predominantly interested not with the 'muscular uterus but with the static and solid structure of the pelvis'.[68] It is the man-midwife who labours, forcing the maternal machine to work efficiently through heroic intervention and the use of instruments. Such representations, it is argued, shaped actual practice, and continue to do so: the 'metaphor of the female body as a defective machine . . . eventually formed the philosophical foundation of modern obstetrics'.[69] Ludmilla Jordanova finds a 'form of implicit violence' in Smellie's illustration of the use of forceps 'as artificial Hands'; 'representational violence', she warns, ultimately acts to 'permit, legitimate, and even encourage actual abuse'.[70] Mechanistic pedagogy and practices govern the experience of the real woman in labour, reducing her to the condition of a clockwork uterus working to the obstetrician's timetable. The degree to which such criticism is justified is arguable: Smellie's writings emphasized rather than undermined uterine force, and he refuted the traditional view that the child is an active participant in labour, showing that the mechanism of parturition turns and expels the child, who is an entirely passive actor.[71] What is intriguing is the pertinence of commentary on Smellie's mechanical and mechanized mothers to the automaton-mother in *Perils of Woman*. The 'forlorn object' is 'delivered of' her child 'in due time': she is passive, a mechanical uterus lacking subjectivity, working (like clockwork) to the established timetable.[72] Post-partum, subjectivity passes to the machine's child, and 'the thing' disappears:

> He was the darling and delight of all concerned with him, while she that gave him birth became as a thing altogether forgotten. [. . .] She was as a thing that had been – that still continued to be, and yet was not![73]

In Gatty Bell, Hogg imagines the inert, mechanical mother detected by Smellie's critics. Gatty is transformed by the automaton from a nervous, morbid girl – a dangerous razor – to a bewildered, submissive mother, reliant for information about her own body on her husband, the surgeon. Her family agree that a 'total change for the better had taken place in her constitution, as well as her intellectual perceptions'.[74] The automaton now seems a reasonable price to pay for this 'sleek, plump' Gatty.[75] M'Ion's retroactive assessment of her ordeal effaces the disgust he felt towards the ghastly automaton, recasting it as a benign body undergoing renovation; but in order to assert this interpretation, he has to stigmatize Gatty's original condition as pathological:

> After a while, the body revived, in the same way as a vegetable revives, but the spirit was wanting; and in that state of healthful and moveless lethargy, have you remained for the long space of three years, unknowing and unknown. At the third return of that momentous day, and on the very hour, the living ray of the divinity returned to enlighten a frame renovated in health, and mellowed to ripeness in all its natural functions, which before were overheated and irrestrainable.[76]

Meiko O'Halloran makes the important point that M'Ion claims 'not that [Gatty] is now herself, but that she is a *better* self'.[77] The qualities of the automaton reverberate back to the body of the young wife, just as the qualities of the automaton of midwifery literature reverberate back to the body of the patient. From razor to the spirit of seasonal revivification, Gatty's body has undergone a hermeneutic migration from monstrosity to sublimity.

M'Ion's hagiography of the automaton is, as O'Halloran observes, both 'overly complacent' and overwhelmed by Gatty's fraught last speech:[78]

> 'I know not what to believe, or what to doubt,' cried she wildly. 'Where have I been? Or rather, *what* have I been? Have I been in a sleep for three years and a day? Have I been in the grave? Or in a madhouse? Or in the land of spirits? Or have I been lying in a state of total insensibility, dead to all the issues of life? What sins may I not have committed during three years of total oblivion?'[79]

M'Ion assures Gatty of her innocence, but withholds the terrible affirmatives to her other questions – that she *was* dressed for the grave, *was* insensible, and *did* sleep in a madhouse. Gatty is never told her full story. The automaton was incoherent, its experience expressed only through physical symptoms of trauma – contortion, spasm, inchoate sound. The task of interpreting these symptoms passed from Gatty to the 'owners' of her body, her husband and the principal physician of the asylum. Both produce inadequate narratives: M'Ion's recuperative reading attempts to turn the automaton into a kind of Sleeping Beauty; the physician's reports confirm only Gatty's continued 'bodily health'.[80] Their calibration of health and sickness is patently at odds with the totality of Gatty's experience. Her pregnancy may be ideal according to the criteria of midwifery literature, in that she is passive, inert and feels no pain; she may emerge from automatism beautiful, sensible and socialized; but these 'improvements' cannot efface the memory of the 'ghastly automaton'. It is the automaton, not the perfected laird's wife, that returns in later scenes of maternal trauma in the concluding parts of the first and third perils. The recuperative reading fails to contain the automaton.

The automaton's legacy

In the perilous history of Scotland described in Hogg's novel, insensibility is perhaps less painful for women than sentience. *Perils of Woman* describes a society that is inherently toxic to women, and particularly to mothers, who exist on the margins of society, on the verge of madness, haunted by echoes of the asylum, their children lost and stolen. The novel's other pregnant women, Katie Rickleton and Sally Niven, seem doomed to repeat Gatty's automatous degradation. Hogg hints at a barely suppressed systemic loathing of reproduction, as when Daniel – thinking his daughter is pregnant with M'Ion's illegitimate child – turns his anger on his flock, declaring 'I'm tired o' thae breeding creatures . . . I shall thin them for aince' and keep only the toops (male sheep).[81] Daniel's fantasy of a world without women (specifically, without gravid women) extends even to the animal kingdom. It is a comic episode, yet its twin conceits – male ownership of female bodies, and male suspicion of female fertility ('breed–breeding', to use Daniel's phrase) – are at work in all three perils.[82] Nor is Daniel's angry leap from human breed–breeding to pastoral management so eccentric, for behind *Perils of Woman's* frequent references to improvement and utility stand the Highland Clearances. Daniel's plan to 'improve' M'Ion's Highland estate by stocking it with sheep complements the 'improvement' of Gatty, mistress of the estate; in both cases, improvement is perceived by those experiencing it in a radically different way to those commenting from afar. Daniel's pastoral ideal becomes terrible reality in the concluding peril's decimation of rural Scotland by war. In the first peril, 'Duff', Daniel's prize toop, is destined to repopulate M'Ion's estate with an improved breed of sheep; in the third peril, the gravedigger 'Davie Duff' is retained to conceal the evidence of the massacre at Culloden and the butchery of the rural population. It is another instance of Hogg's brutal irony: Davie clears the land of the corpses of its original inhabitants, and in the future, his namesake restocks it with more productive flesh. The direction of society is towards reduction, even as it demands reproduction from female bodies. Promising proliferation and continuity (the antithesis to clearance and improvement), the maternal body buckles under the contesting pressures to reproduce and reduce.

Katie's story is a comic reworking of Gatty's tale. She marries Richard Rickleton when pregnant with another man's child, a scheme first attempted by Gatty's parents when they thought she had been seduced by M'Ion. Refusing to give up her illegitimate child, Katie is threatened with the apparatus of the asylum by her doctor, seducer, mother and nurse, who resolve to 'take the child from her by force, even though it should be found necessary to put her in a straightjacket, and bind both her hands and her feet'.[83] Katie is rescued from this brutal re-enactment of Gatty's objectification and incarceration by her husband, who unlike M'Ion – and in a remarkable departure from 'polite' morality – does *not* exile his degenerate wife. Though Katie teeters on the edge of subjection, offering to become her husband's 'slave' in gratitude, Rickleton refuses to demean her, declaring her to be 'the lady of my right hand' and adopting her child.[84] If any story in *Perils of Woman* can be said to have a happy ending, it is the Rickletons', which resolves the 'problem' of an undisciplined maternal body by rejecting the moral codes which render it problematic, thereby enfranchizing an illegitimate child who, as the adopted emblem of an Anglo-Scottish union, is perhaps a more significant emblem of national unity than Gatty's son, the heir to land enriched by the Highland Clearances.

In reading order, however, *Perils of Woman* concludes not with the joyful moral elasticity of the Rickletons, but with the tragedy of Sally Niven, whose demise and revivification as an animated corpse is 'the narrative issue' of the 'scandalous reduction of the domestic national heroine' in the first peril.[85] Like Gatty, Sally is a lowlander married to a highlander, for whom maternity brings insanity and living death. No sentimental heroine, Sally is witty and wily, but proves as vulnerable as Gatty to the pressure of irresistible forces. Her temporary resurrection repeats Gatty's, 'detail for detail'.[86] A male hand reanimates the (apparently) dead body of a pregnant woman, which is actually a living, mad, violent body. Sally's detachment from reality is, like Gatty's, expressed through disconnected speech. Traumatized by the murder of her husband and protector, she speaks with 'an incoherence of metaphor, and allusions, that a healthful mind would scarcely have framed'.[87] In the devastating conclusion, she is a mad mother singing to her dead daughter in the wilderness, having fled the 'asylum' (a loaded term, given Gatty's fate) provided by sympathizers.[88] Though chronologically Sally's tragedy precedes Gatty's tale, in reading order it reiterates Gatty's degradation, so that the piling up of repetitions creates a feeling of dreadful inevitability – Sally's pregnancy promises not life and reconciliation, but psychological disintegration, death and the rupture of history. The automaton's legacy is, it seems, inescapable. Though their fates are antithetical (Sally dies, Gatty is perfected), the striking parallelism of their degradation reveals the thesis in the antithesis, so that Sally's degeneration counterpoints Gatty's supposed improvement.

Ostensibly commending the automaton as a maternal ideal, *Perils of Woman* is in fact unremittingly sceptical of reductive practices and interpretations. Gatty's efficient automatism generates a child and a renovated matriarch, but at the price of degradation. Her distress renders uncertain the unification achieved by the automaton, a suspicion that is confirmed by the death of Sally and her daughter, which extinguishes the hopes of the second and third peril for unification and survival. But Gatty's and Sally's dreadful doubling encircles a solitary subversive survivor. Katie Rickleton, the transgressor, delivers and saves the child who – as Richard Rickleton's adopted son – will be heir to a fully formed, genuine social community. Katie's nested tale ends hopefully with a series of reconciliations and reunions, between Katie and Richard, and thus between Lowland Scotland and Northern England, and between Richard and his inveterate enemy, who petitions to sponsor the child's baptism. 'I AM HAPPY,' Richard writes defiantly, and so ends his 'sublime remonstrance on the impropriety of breaking the Seventh Commandment, especially on the part of the women'.[89] Katie's redemption from automatous subjugation is secreted at the centre of the novel's structure of nested narratives, making the transgressive mother, who resisted the paraphernalia of the asylum and escaped subjection, the real heart of the novel. With such a subversive conclusion to such a tale, in such a bewildering novel, it is unsurprising that *The Three Perils of Woman* was not resurrected for a Victorian readership.

Notes

1. James Hogg, *The Three Perils of Woman, or Love, Leasing, and Jealousy, a series of Domestic Scottish Tales*, ed. by Antony Hasler and Douglas Mack (Edinburgh: Edinburgh University Press, 2002). All further references are to this edition.
2. On the critical reception of *Perils of Woman*, see David Groves, 'Afterword', in James Hogg, *The Three Perils of Woman, or Love, Leasing, and Jealousy, a series of Domestic Scottish Tales*, ed. by Antony Hasler and Douglas Mack (Edinburgh: Edinburgh University Press, 1995), 409–37. The Afterword appears only in this edition.
3. John Barrell, 'Putting Down the Rising', in Leith Davis, Ian Duncan and Janet Sorensen (eds), *Scotland and the Borders of Romanticism* (Cambridge: Cambridge University Press, 2004), 130.
4. On Hogg's subversive reinterpretation of the national tale heroine, see Antony J. Hasler, 'Introduction', in Hogg, *The Three Perils of Woman* (2002), and Ian Duncan, *Scott's Shadow: The Novel in Romantic Edinburgh* (Princeton: Princeton University Press, 2007), 208–12.
5. Hogg, *The Three Perils of Woman*, 201.
6. Sigmund Freud, 'The Uncanny', in Albert Dickson (ed.), *Art and Literature* (London: Penguin, 1990), 345.
7. See *Dictionary of the Scots Language*, www.dsl.ac.uk
8. Freud, 'The Uncanny', 347.
9. Steven Connor, *Dumbstruck: A Cultural History of Ventriloquism* (Oxford: Oxford University Press, 2000), 340–41.
10. David Brewster, *Edinburgh Encyclopaedia* (Edinburgh: Blackwood, 1830; repr., London: Routledge, 1999), 3:151.
11. Brewster, *Edinburgh Encyclopaedia*, 2:62.
12. Hogg, *The Three Perils of Woman*, 182.
13. Hogg, *The Three Perils of Woman*, 182.
14. Hogg, *The Three Perils of Woman*, 182–83.
15. Hogg, *The Three Perils of Woman*, 224.
16. Hogg, *The Three Perils of Woman*, 193.
17. A range of explanations for Gatty's illness have been proposed, including venereal disease, catatonic schizophrenia, hysteria and divine retribution.

See David Groves, 'James Hogg's *Confessions* and *The Three Perils of Woman* and the Edinburgh Prostitution Scandal of 1823', *Wordsworth Circle*, 18 (1987), 127–31; '*The Three Perils of Woman* and the Edinburgh Prostitution Scandal of 1823', *Studies in Hogg and his World* 2 (1991), 95–102; and 'Afterword' in Hogg, *The Three Perils of a Woman* (1995); Barbara Bloedé, 'Hogg and the Edinburgh Prostitution Scandal', *Newsletter of the James Hogg Society*, 8 (1989), 15–18; and Barbara Bloedé, '*The Three Perils of Woman* and the Edinburgh Prostitution Scandal: A Reply to Dr Groves', *Studies in Hogg and His World*, 3 (1992), 88–94.

18. Duncan, *Scott's Shadow*, 210.

19. James Hogg, *Anecdotes of Scott*, ed. by Jill Rubenstein (Edinburgh: Edinburgh University Press, 2004), 61; and Duncan, *Scott's Shadow*, 208.

20. Hogg, *The Three Perils of Woman*, 199.

21. Duncan, *Scott's Shadow*, 209.

22. Hogg, *The Three Perils of Woman*, 203.

23. John Aldini, *General Views on the Application of Galvanism to Medical Purposes; Principally in Cases of Suspended Animation* (London: Callow, 1819), 26.

24. Duncan, *Scott's Shadow*, 210.

25. Gatty's reanimation has previously been discussed in relation to Aldini's experiments: see Duncan, *Scott's Shadow*, 209; and Richard D. Jackson, 'Gatty Bell's Illness in James Hogg's *The Three Perils of Woman*', *Studies in Hogg and his World*, 14 (2003), 16–29.

26. On the stylistic scission of the version in Ure's *Dictionary of Chemistry* (1821), see Charlotte Sleigh, 'Life, Death and Galvanism', *Studies in History and Philosophy of Biological and Biomedical Sciences*, 29 (1998), 243–44.

27. Andrew Ure, 'An Account of Some Experiments Made on the Body of a Criminal Immediately after Execution, with Physiological and Practional Observations', *Journal of Science and the Arts*, 6 (1819), 290.

28. Ure, 'An Account of Some Experiments', 292.

29. Hogg, *The Three Perils of Woman*, 199–200.

30. Ure, 'An Account of Some Experiments', 289.

31. Sleigh, 'Life, Death and Galvanism', 243.

32. Hogg, *The Three Perils of Woman*, 201.

33. Hogg, *The Three Perils of Woman*, 201.

34. Hogg, *The Three Perils of Woman*, 201.

35. Hogg, *The Three Perils of Woman*, 202–4, my emphasis.

36. Hogg, *The Three Perils of Woman*, 202, my emphasis.

37. Hogg, *The Three Perils of Woman*, 205.

38. Antony J. Hasler, '*The Three Perils of Woman* and John Wilson's *Lights and Shadows of Scottish Life*', *Studies in Hogg and his World*, 1 (1990), 33.

39. Barrell, 'Putting Down the Rising', 131; and Douglas Mack, 'Hogg and Angels', *Studies in Hogg and his World*, 15 (2004), 21.

40. Hogg, *The Three Perils of Woman*, 224.

41. Hogg, *The Three Perils of Woman*, 224.

42. Mack, 'Hogg and Angels', 94.

43. Hogg, *The Three Perils of Woman*, 206.

44. Hogg, *The Three Perils of Woman*, 204.

45. Duncan, *Scott's Shadow*, 211.

46. Lisa Forman Cody, *Birthing the Nation: Sex, Science, and the Conception of Eighteenth-Century Britons* (Oxford: Oxford University Press, 2005), 27.
47. Duncan, *Scott's Shadow*, 210.
48. Gillian Hughes, *James Hogg: A Life* (Edinburgh: Edinburgh University Press, 2007), 173.
49. John Thatcher, *A Letter to the Right Honourable the Lord Provost and Patrons of the University of Edinburgh, On the Proposed New Regulation Respecting the Study of Midwifery* (Edinburgh: Carfrae, 1825), 6. Professionalization in fact progressed slowly. Professor James Hamilton was not admitted to the Medical Faculty until 1830, and even then, the diploma in midwifery was issued separately from the graduation diploma. See A.R. Simpson, 'History of the Chair of Midwifery and the Diseases of Women and Children in the University of Edinburgh', *Edinburgh Medical Journal*, 28 (1882), 496.
50. Ornella Moscucci, *The Science of Woman: Gynaecology and Gender in England, 1800–1929* (Cambridge: Cambridge University Press, 1990), 48–49.
51. Andrea K. Henderson, *Romantic Identities: Varieties of Subjectivity 1774–1830* (Cambridge: Cambridge University Press, 1996), 12.
52. Bryan Hibbard, *The Obstetrician's Armamentarium: Historical Obstetric Instruments and Their Inventors* (San Anselmo: Norman, 2000), 79.
53. Hibbard, *The Obstetrican's Armamentarium*, 80.
54. R.W. Johnstone, *William Smellie: The Master of British Midwifery* (Edinburgh: E&S Livingstone, 1952), 26.
55. Samuel Paterson, *A Catalogue of the Entire and Inestimable Apparatus for Lectures in Midwifery, Contrived with Consummate Judgment, and Executed with Infinite Labour, by the Late Ingenious Dr. William Smellie, Deceased: Consisting of a Variety of Anatomical Preparations, Illustrating the Theory of Midwifery, the Original Drawings by Rymsdyk, from which his Engravings were Made, his Exquisite Artificial Machines, in Imitation of the Living Subjects, his Collection of Obstetrical Instruments, English and Foreign* (London: Royal College of Obstetricians and Gynaecologists, 1770), 6.
56. Paterson, *A Catalogue*, 6.
57. William Smellie, *A Course of Lectures upon Midwifery, wherein the Theory and Practice of that Art are Explain'd in the Clearest Manner. More particularly, the Structure of the Pelvis and Uterus. Of the Foetus in Utero, and after Parturition. The Management of Child-Bearing Women, during Pregnancy, in time of Labour, and after Delivery. The Manner of Delivering Women, in all the Variety of Natural, Difficult, and Preternatural Labours, Perform'd on Different Machines made in Imitation of Real Women and Children* (London, 1745), Wellcome Library MS 4630.
58. William Smellie, *A Course of Lectures upon Midwifery*, 5.
59. William Smellie, *Smellie's Treatise on the Theory and Practice of Midwifery* (London: The New Syndenham Society, 1876), 25.
60. Smellie, *A Course of Lectures upon Midwifery*, 44.
61. T. Percy Kirkpatrick, *The Book of the Rotunda Hospital: An Illustrated History of the Dublin Lying-In Hospital From its Foundation in 1745 to the Present Time* (London: Bartholomew, 1913), 81–82.
62. Cody, *Birthing the Nation*, 169.
63. Pam Lieske, 'William Smellie's Use of Obstetrical Machines and the Poor', *Studies in Eighteenth-Century Culture*, 29 (2000), 65–86.

64. Bonnie Blackwell, '*Tristram Shandy* and the Theater of the Mechanical Mother', *ELH*, 68 (2001), 92–93, 127.
65. Henderson, *Romantic Identities*, 16.
66. William Smellie, 'Preface' to *A Sett of Anatomical Tables, With Explanations, and an Abridgment, of the Practice of Midwifery, With a View to Illustrate a Treatise On that Subject, and Collection of Cases* (London, 1754).
67. William Hunter, 'Preface' to *The Anatomy of the Gravid Uterus Exhibited in Figures* (Birmingham: Baskerville, 1774).
68. Henderson, *Romantic Identities*, 14–15.
69. Robbie E. Davis-Floyd, 'The Technocratic Model of Birth', in Philip K. Wilson (ed.), *The Medicalization of Obstetrics: Personnel, Practice and Instruments* (New York and London: Garland, 1996), 251.
70. Ludmilla Jordanova, *Sexual Visions: Images of Gender in Science and Medicine between the Eighteenth and Twentieth Centuries* (Hemel Hempstead: Harvester Wheatsheaf, 1989), 61–62.
71. John Glaister, *Dr. William Smellie and his Contemporaries: A Contribution to the History of Midwifery in the Eighteenth Century* (Glasgow: Maclehose, 1894), 192.
72. Hogg, *The Three Perils of Woman*, 204.
73. Hogg, *The Three Perils of Woman*, 205.
74. Hogg, *The Three Perils of Woman*, 215.
75. Hogg, *The Three Perils of Woman*, 213.
76. Hogg, *The Three Perils of Woman*, 224.
77. Meiko O'Halloran, 'Treading the Borders of Fiction: Veracity, Identity and Corporeality in *The Three Perils*', *Studies in Hogg and his World*, 12 (2001), 48–49.
78. O'Halloran, 'Treading the Borders of Fiction', 48–9.
79. Hogg, *The Three Perils of Woman*, 223.
80. Hogg, *The Three Perils of Woman*, 203.
81. Hogg, *The Three Perils of Woman*, 117.
82. Hogg, *The Three Perils of Woman*, 117.
83. Hogg, *The Three Perils of Woman*, 239.
84. Hogg, *The Three Perils of Woman*, 255.
85. Duncan, *Scott's Shadow*, 212.
86. Barrell, 'Putting Down the Rising', 134.
87. Hogg, *The Three Perils of Woman*, 398.
88. Hogg, *The Three Perils of Woman*, 406–7.
89. Hogg, *The Three Perils of Woman*, 257–8.

4
Clockwork Automata, Artificial Intelligence and Why the Body of the Author Matters

Paul Crosthwaite

On 9 June 1949, Geoffrey Jefferson, Professor of Neurosurgery at the University of Manchester, marked his receipt of the prestigious Lister Medal from the Royal College of Surgeons by addressing the College's members on the topic of 'The Mind of Mechanical Man'. Jefferson summarized his assessment of the prospects for mechanical consciousness with these words:

> Not until a machine can write a sonnet or compose a concerto because of thoughts and emotions felt, and not by the chance fall of symbols, could we agree that machine equals brain – that is, not only write it but know that it had written it. No mechanism could feel (and not merely artificially signal, an easy contrivance) pleasure at its successes, grief when its valves fuse, be warmed by flattery, be made miserable by its mistakes, be charmed by sex, be angry or depressed when it cannot get what it wants.[1]

This passage articulates a set of principles that would come to guide much of the research in the fields that we now know as artificial intelligence (AI) and artificial life (AL): that there is a fundamental difference between the mere rote processing of data and the self-conscious awareness and understanding of what is being processed; that authentic consciousness is coloured by shifting emotional, affective and libidinal states; and that this rich psychic reality finds its privileged expression in acts of artistic creation. This essay explores how these conceptions have been channelled into attempts to design computer programmes capable of producing original works of literature. It does so, however, by drawing parallels between recent research by computer scientists into the possibility of constructing artificial authors and the elaborate clockwork writing automata produced by European craftsmen in the late eighteenth to mid-nineteenth centuries. Building on the work of Jessica Riskin and others, I suggest that for all their comparative lack of sophistication, the clockwork writers of the eighteenth and nineteenth centuries anticipate – by virtue of their embodiment in articulated, humanoid form – an emerging paradigm in the field of artificial intelligence that presents tantalizing possibilities for the development of machine creativity. This new research agenda has, in turn, surprising and profound consequences for literary criticism and theory today.

Writing automata and romantic authorship

Between 1768 and 1844, the Swiss watchmakers Pierre and Henri-Louis Jaquet-Droz and Jean-Frédéric Leschot, the Jaquet-Drozes' protégé Henri Maillardet, and the French illusionist Jean-Eugène Robert-Houdin constructed a series of startlingly lifelike writer-figures (figures 5–8).

Seated at their desks and equipped with quills, they astounded viewers by inscribing documents set before them with signatures or short poems.

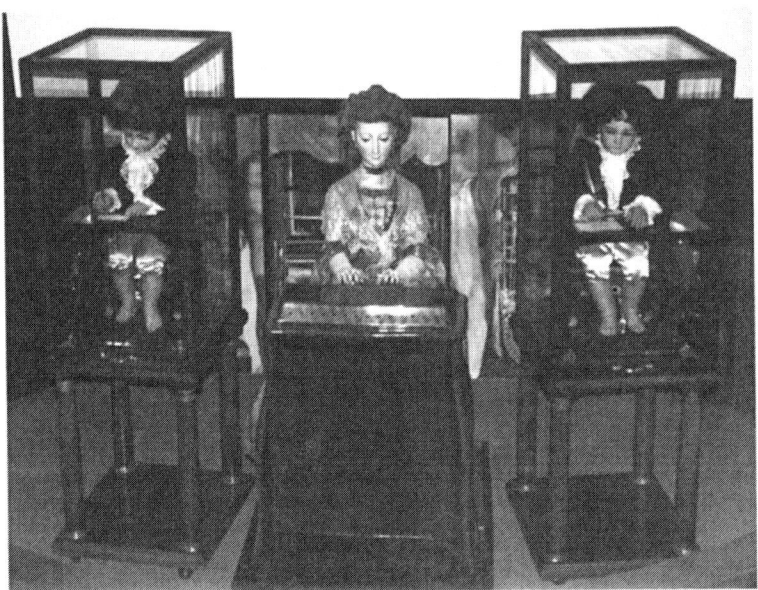

Figure 5: 'The Draftsman', 'The Musician' and 'The Writer', constructed 1768–1774, by Pierre and Henri-Louise Jaquet-Droz and Jean-Frédéric Leschot

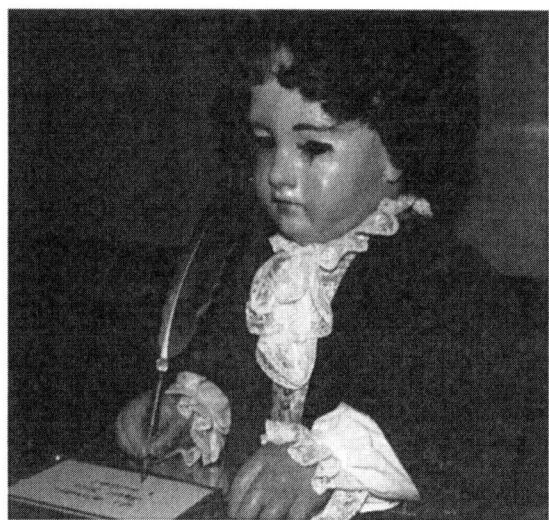

Figure 6: 'The Writer' by Pierre and Henri-Louise Jaquet-Droz and Jean-Frédéric Leschot

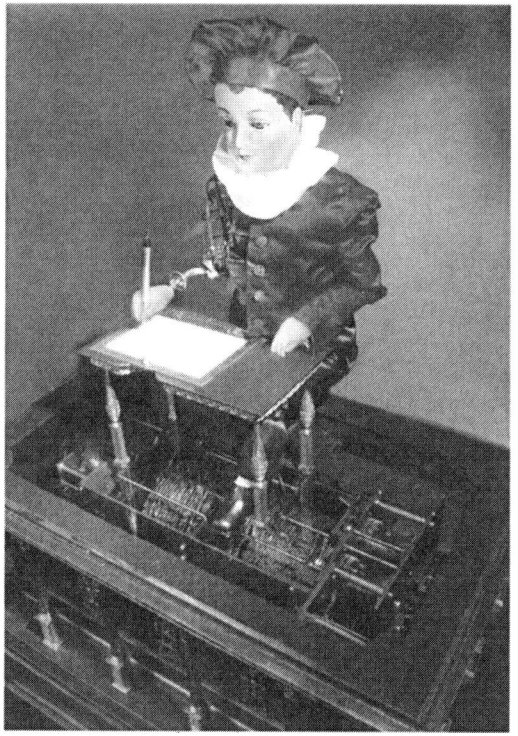

Figure 7: 'Writer-Draughtsman' (1805) by Henri Maillardet

Figure 8: Jean-Eugène Robert-Houdin with his Writer-Draughtsman (1844; destroyed by fire 1865) in background

The texts to be written by the Jaquet-Droz and Maillardet devices were 'coded' by selecting letters on a wheel and/or by setting a series of cams. Turned by clockwork motors, these components drove complex systems of levers and rods, which guided the movements of the figures' hands over the page. Robert-Houdin's writer (Figure 8) is thought to have been destroyed by a fire in 1865, and so has not been subjected to expert examination, but it seems likely that, in common with other automata built by the great illusionist, it would have combined elements of the sophisticated mechanisms utilized by the Jaquet-Drozes and Maillardet with hidden levers or pedals controlled by a human operator.[2] Jessica Riskin locates this difference in the construction of these automata in the context of a shift from an eighteenth-century ethos of simulation (which sought to replicate, as accurately as possible, the mechanics of physiological processes themselves) to a nineteenth-century culture of analogy (which was content with devices that merely presented an outward semblance of such underlying processes).[3] As Riskin acknowledges, however, even the device she identifies as the principal embodiment of the philosophy of simulation – Jacques Vaucanson's digesting duck (first exhibited 1738) – employed a crude fraud, rather than the elaborate technological architecture proclaimed by its maker, to achieve its effects (a fraud which was eventually revealed by none other than Robert-Houdin).[4] Without discarding Riskin's distinction between simulation and analogy, then, it is nonetheless evident that throughout the eighteenth and early nineteenth centuries, automaton makers were united in manipulating the outwardly visible form and function of their machines in order to effect an illusion of interior organic process, irrespective, to some degree at least, of the actual process employed. In the case of the writing automata of the period, the intended illusion, of course, was the uncanny impression that the figures were not simply rehearsing a series of predetermined actions, but independently generating their own writings. Self-evidently, moreover, this illusion was reliant on the convincingly humanoid appearance and behaviour of the writers, while it would have been dispelled by the exposure of the mechanism (of whatever kind) propelling their motions. As a character in E.T.A. Hoffmann's 'Die Automaten' (1821) says of an automaton he has observed: 'The outward form . . . [of the figure] has been cleverly selected. Its shape, appearance and movements are well adapted to occupy our attention in such a manner that its secrets are preserved and to give us a favourable opinion of the intelligence which gives the answers'.[5] The impression of autonomous agency thus conveyed could be quite profound. John Tresch has recently charted the varied and ambiguous responses that such devices elicited when they were displayed to the public in theatres and exhibition halls. While some, generally more educated, viewers approached them as amusing novelties or impressive displays of technical ingenuity, others were willing to entertain the notion that they possessed some genuine flash of vitality.[6] It is striking, though, that in accounts of public displays of these figures, even the most sober witnesses testify to at least a fleeting illusion of spontaneous creativity. As I have suggested, this effect was dependent on the machines' status as androids (humanoid automata) with anatomically proportionate physiques and naturalistic physiognomies; and this was for two reasons, both of which bear on prevailing models of subjectivity and authorship in the period.

Much discussion surrounds the cultural, philosophical and aesthetic meanings and associations of these and other, contemporaneous, automata. Simon Schaffer, for example, identifies them as literal embodiments of a tradition of seventeenth- and eighteenth-century European thought that viewed humans as purely mechanical beings, a tradition whose origins lie in René Descartes' philosophy of animal existence, and which, in extending Descartes' claims to humans, reached its culmination in the French Enlightenment materialist ideas of Denis Diderot and, pre-eminently, Julian Offray de La Mettrie. Its definitive statement is La Mettrie's 1747 treatise *Machine Man*, with its conclusion that 'man is a machine and . . . there is in the whole universe only one diversely modified substance'.[7] As Riskin emphasizes, though, even the most mechanistic of French Enlightenment philosophies found no contradiction in celebrating humanity's capacity for 'sentiment' and 'sensibility' – for feeling, emotion, passion and expression;[8] the attempts of automaton makers to employ mechanical means in order to build figures that exhibited all the animation and vitality of human beings was, she suggests, wholly consistent with this outlook. The importance of a language of sentiment and sensibility throughout the eighteenth and early nineteenth centuries is, furthermore, just one of many continuities that have led critics and cultural historians to question conventional divisions between Enlightenment and Romantic world-views.[9] Over a period exactly contemporaneous with the production of the clockwork writers discussed above, this tradition effloresced into the Romantic discourse of authorship, with its vision of literature as 'fundamentally expressive of a unique individuality' and defined by 'originality' and the 'conscious intention of the autonomous subject'.[10] Just as Romanticism privileged the expressive capacity of the writer, so, as Christopher Keep argues, the ability of an automaton to write suggests, more strongly than any other function it might possess,

> the presence not of a program but of a person, one whose actions are the free and spontaneous expressions of some deep reserve of selfhood, an inwardness or depth of being which is capable of reflecting on itself as self. The very appearance of writing . . . is always marked by the trace or outline of a living presence, the unique individual who is both the source and origin of the enunciative act.[11]

Vivian Sobchack also interprets these machines, like similar devices now marketed as children's toys, as dramatizing a Romantic conception of authorship through the act of writing – specifically through the act of writing *by hand*. Handwriting, she remarks, 'is always . . . *auratic* insofar as it is enabled not just by a material body but by a *lived body* that, however regulated, cannot avoid inscribing its singular intentionality in acts and marks of *expressive improvisation*'.[12]

The author of Romantic theory is a notoriously contradictory being, however, and if the convincingly anthropomorphous performances of these mechanical writer-figures partake of an expressivist or idealist vision of literary creativity, in which the imagination is granted an autopoietic status, they equally resonate with an empiricist model that stresses the writer's constitutive receptivity to the dynamics of the external world. As Riskin notes, the automaton makers of the eighteenth and nineteenth centuries did not attempt to replicate the action of the five senses;[13] again, though, the verisimilar appearance of their machines succeeded in conveying to viewers an impression of responsiveness and alertness. In the case of the Jaquet-Drozes, even the choice of the figure's footwear (or lack of it) was designed to suggest an acute sensitivity to the environment (see Figure 5). As Gaby Wood puts it,

> some inventors intended their objects to be artificial forms of an eighteenth-century ideal – the child as a blank slate, the purest being. The Jaquet-Droz figures conduct their marvellous activities barefoot, illustrating a belief, held by their contemporary Jean-Jacques Rousseau, that children would learn more freely if unhampered by shoes.[14]

As has often been noted, the exemplary synthesis of the expressivist-idealist and empiricist currents in Romantic aesthetics is found in William Wordsworth's 'Tintern Abbey' (1798).[15] Wordsworth had no enthusiasm for automata,[16] but his exultant celebration of 'all the mighty world / Of eye and ear, both what they half-create, / And what perceive' (106–08) precisely delineates the spectrum of faculties that the clockwork writers of his day were designed to give the impression of possessing. I will return to Wordsworth's poem later, in light of the connections I now wish to draw between these writer-figures and some recent developments in artificial intelligence research. If, in the late eighteenth and early nineteenth centuries, the writing automaton's humanoid and (partially) mobile construction was technically incidental, and served merely to generate an illusion of sentience and perception, in our own era such material embodiment is increasingly viewed as essential to the creation of genuine artificial intelligence.

Meaning and embodiment in machine-generated literature

In November 1928, the Franklin Museum in Philadelphia took delivery of the damaged and disassembled components of a brass clockwork machine. The donors, in whose family the device had resided for several generations, understood it to have once been capable of writing messages and drawing pictures, and had some notion of it being the work of the German inventor Johann Nepomuk Maelzel. After an engineer at the Institute had painstakingly repaired the device, it was equipped with a fountain pen and set in motion. It promptly inscribed four drawings and three poems, signing the last with the flourish, '*Ecrit par L'Automate de Maillardet*'.[17] This wonderfully eerie story – a signal manifestation of that effect we have come to call the uncanny – perfectly allegorizes the historical dynamic I wish to explore, in which the fidelity of later generations to the embodied form of early clockwork automata permits those figures to address us, with arresting directness, across the centuries.

Riskin notes that the conviction in this earlier moment 'that life, consciousness and thought were essentially embodied in animal and human machinery has striking parallels in current Artificial Intelligence'.[18] As she observes, the notion that intelligence must be 'physically grounded' is the central principle of the sub-discipline of AI known as artificial life (AL). She cites the pioneering work of Rodney Brooks, director of the Artificial Intelligence Lab at MIT, who 'has left behind the purely software model of AI, and instead builds robots with sensors and feedback loops, giving them vision, hearing and touch'.[19] The idea that, as Susan Blackmore puts it, 'mind can be created only by interacting in real time with a real environment' is of particular significance in the branches of AI that attempt to equip machines with a grasp of language.[20] Perhaps the most significant challenge to conventional, box-bound AI programmes in this regard is the argument – made, most influentially, by the philosopher of mind John Searle – that such systems will never possess genuine linguistic ability because they are condemned to an existence that lacks 'intentionality'; that is, while they may be able to follow (and even, with the advent in the 1980s of artificial neural networks, progressively learn or internalize) syntactic rules, they can have no understanding of what the symbols they manipulate according to these rules are 'about', what they signify or refer to, what they *mean*. The 'bottom-up' approach advocated by proponents of 'embodied cognition' seeks to redress the problem of intentionality by more closely replicating the processes of human language acquisition:

> As human infants develop linguistic competence, they learn not only how to describe objects but also how to describe and express intentional relations such as wants, likes, and dislikes, intentional

relations that were experienced by the infant before they could be cast in linguistic form. In this way language and cognition elaborate on previously experienced nonlinguistic and noncognitive (i.e., bodily) intentional relationships. An infant's experiences of noncognitive intentional relationships provide the foundation necessary for the cognitive life it will later enjoy. Human beings do not suffer from the symbol grounding problem . . . precisely because we are embodied. [. . .]

By building robots that interact with the environment prior to linguistic competence . . . an embodied approach to AI provides the necessary foundation for higher cognition.[21]

This new paradigm has not so far been extended into the domain of machine-generated literature. The potential of embodied cognition for this field can be most readily grasped, however, by considering the software-based systems that currently exist. I will focus on the two most prominent examples: the Cybernetic Poet designed by the American inventor Ray Kurzweil, and BRUTUS, a short-story-writing programme developed by Selmer Bringsjord, David Ferrucci and a team of computer scientists at Rensselaer Polytechnic Institute in New York State. Given their significance for questions of authorship and textual meaning, and the considerable scholarly interest in other forms of electronic literature such as hypertext and interactive fiction, these projects have received remarkably little attention in literary studies.[22]

Kurzweil's Cybernetic Poet works by 'reading' – as its inventor puts it – poems by an author or authors and constructing a model of their work. The programme develops algorithms that allow it to imitate the style, metre, poetic structure and vocabulary characteristic of the author(s). Kurzweil states that the resulting 'poems are in a similar style to the author(s) originally analyzed but are completely original new poetry'.[23] The following are examples of poems generated by Kurzweil's programme, which can be accessed on his website:

Soul

A haiku written after reading poems by John Keats and Wendy Dennis

You broke my soul
the juice of eternity,
the spirit of my lips.

And Pink In Sex

A haiku written after reading poems by Walt Whitman

Ages and pink in Sex,
Offspring of the
voices of all my Body.

Bringsjord and Ferrucci's BRUTUS produces short stories of up to 500 words using a sophisticated 'story grammar' architecture, which can handle character, setting, plot development and the other basic elements of prose narrative. The first incarnation of BRUTUS took eight years to develop. Its designers concentrated on equipping the programme with the ability to write stories centred around the theme of betrayal, since this was one aspect of human experience that, they reasoned, could be logically tabulated, in contrast to more diffuse emotional phenomena such as love, fear or regret. One of BRUTUS's stories, entitled simply 'Betrayal', begins like this:

> Dave Striver loved the university. He loved its ivy-covered clocktowers, its ancient and sturdy brick, and its sun-splashed verdant greens and eager youth. He also loved the fact that the university is free of the stark unforgiving trials of the business world – only this *isn't* a fact: academia has its own tests, and some are as merciless as any in the marketplace. A prime example is the dissertation defense: to earn the PhD, to become a doctor, one must pass an oral examination on one's dissertation. This was a test Professor Edward Hart enjoyed giving.[24]

Literature generators such as the Cybernetic Poet and BRUTUS invite comparison with an avant-garde tradition of 'machine writing' that extends back to the beginning of the twentieth century. As Brian McHale demonstrates in an important recent survey, this tradition encompasses such varied innovations as the procedural compositional techniques employed by Raymond Roussel, the literary 'games' developed by the Surrealists, the cut-up strategies pioneered by the Dadaists and later adopted by William Burroughs, the author-computer collaborations undertaken by Charles O. Hartman and the aleatory and/or arbitrarily rule-bound methods pursued by Louis Zukofsky, Jackson Mac Low, the Language poets and the OuLiPo circle. While only a few of these figures and movements utilize actual machines, their texts are all examples of machine writing in the sense of that they are 'not "freely" composed but produced by the operation of mechanical techniques for generating and/or manipulating bits of language'.[25] McHale argues that these texts can be best compared to one another 'in terms of the relative proportions of writer to machine participation in the composition of the text'.[26] All literature, he argues, possesses a mechanical element to the extent that it imposes constraints of form, genre, length – and, specifically in the case of poetry, rhyme and metre – on the expressive capacity of the author.[27] The spectrum of machine writing as such, however, ranges from texts in which a set of more-or-less mechanically produced materials are submitted to heavy postprocessing on the part of the writer (as in the novels of Raymond Roussel), through to the practice of a figure such as Jackson Mac Low, which entirely delegates the outcome of a particular procedure or programme, and even in certain instances the choice of the procedure itself, to mechanical permutation.[28] Kurzweil's and Bringsjord and Ferrucci's generators would find themselves at the latter pole of McHale's typological scale: they produce their texts within certain pre-programmed parameters, but the process is not otherwise subject to human intervention or interference. While avant-gardist utilizations of mechanical techniques consistently present themselves as strategic subversions of the autonomous, expressive author who looms so large in Romantic and post-Romantic aesthetic ideology, however, the Cybernetic Poet and BRUTUS projects seemingly aspire to endow the machine itself with those very qualities of sovereignty and creative genius. Tellingly, Kurzweil has a clear preference for poets in the Romantic tradition, including Keats and Whitman, as well as Blake, Byron and Shelley (though the relation of his own mechanical poet to these titans is, of course, purely imitative). Bringsjord and Ferrucci, meanwhile, identify the benchmark of creativity with such giants of the canon as Dickens, Tolstoy, Joyce, Updike and Morrison (whilst acknowledging that 'if BRUTUS$_n$, some refined descendant of BRUTUS$_1$, is to soon find employment at the expense of a human writer, in all likelihood it will be as an author of formulaic romance and mystery'[29]).

Regardless of the shortcomings of these programmes with respect to their designers' highest ambitions, and whatever we might think of the aesthetic qualities of the writings they produce, it is undeniable that their texts bear at least a passable resemblance to the literary forms they are designed to emulate. As Kathleen L. Komar says of a piece by Kurzweil's poet, 'if we did not know this [the text's mechanical provenance], we would undoubtedly count the poem as literature'.[30] What is equally clear, however (and what I take to be the grounds for Komar's equivocation), is that these programmes remain entirely bound by the problem of intentionality: they may be able to follow rules in such a way as to produce texts that meet the objective criteria for recognition as works of poetry or short fiction, but they have no apprehension of what these texts mean, or even that they *could* yield such a thing as meaning. The successes and failures of these projects are cast into sharp relief by two celebrated intellectual experiments, which are often hailed as inaugurating, respectively, the disciplines of artificial intelligence and modern literary theory: the eponymous test invented by the computer scientist Alan Turing to determine the existence of machine intelligence, and the experiment in literary response undertaken by the critic I.A. Richards under the banner of 'practical criticism'.

The 'Turing Test', first described in the seminal paper 'Computing Machinery and Intelligence' (1950), consists of a scenario in which a human interrogator poses questions to two concealed interlocutors in an attempt to determine which is a human and which a machine; if the machine can persuade the interrogator that it is the human party, Turing reasoned, then it can be legitimately deemed intelligent. Both the Cybernetic Poet and BRUTUS have been submitted to variations on the Turing Test, in which readers attempted to distinguish the programmes' texts from those by human writers.[31] The thirteen adults and three children to whom Kurzweil administered his test correctly attributed the poems they read at a rate of 63 per cent and 48 per cent, respectively.[32] Meanwhile, 25 per cent of the two thousand web visitors who read a piece by BRUTUS alongside four stories by human writers successfully identified the machine-authored text.[33] The fact that in these (albeit only semi-scientific) tests readers succeeded in distinguishing between human- and computer-generated writings at a rate not significantly better than chance lends empirical weight to the assertion that the two programmes are capable of imitating the conventions of their assigned genres with a considerable degree of credibility. Paradoxically, however, when the products of these programmes are considered in light of an experiment that insists on curbing considerations of authorial identity in favour of concentrated interpretation of the texts themselves, the irrevocable alterity of their computational origins – seemingly elided in these quasi-Turing Tests – reasserts itself.

I.A. Richards' *Practical Criticism* (1929) describes an initiative undertaken at Cambridge University in the 1920s, whereby groups of readers, predominantly undergraduates studying English, were issued with poems by a range of authors – contemporary, canonical and minor – which had been stripped of personally and historically identifying details; the readers were invited to submit written responses or 'protocols' in which they recorded their reflections on these materials. Richards' realization, on the basis of these often misconceived submissions, that literary criticism needed to develop a considerably more rigorous and systematic methodology prepared the ground for many of the theoretical innovations of the succeeding decades. His pedagogical exclusion of biographical and historical data so as to focus the reader's attention on the words on the page would prove to be particularly significant for the Anglo-American New Critics of the 1940s and 1950s. Its influence is notably discernible in William K. Wimsatt and Monroe C. Beardsley's famous refutation of the so-called intentional fallacy on the grounds that 'the design or intention of the author is neither available nor desirable as a standard for judging the success of a work of literary art'.[34] Despite the very different intellectual coordinates of the two traditions, affinities can also be detected between the vision of the text as an autonomous artefact pioneered by Richards and formalized by the New Critics and the radical anti-authorialism and anti-intentionalism of certain strands of French poststructuralist theory, positions most vividly articulated in Roland Barthes' notorious 1967 essay 'The Death of the Author'. There are, for example, clear resonances with Wimsatt and Beardsley in Barthes' claim that in traditional criticism 'the *explanation* of a work is always sought in the man or woman who produced it, as if it were always in the end . . . the voice of a single person, the *author* "confiding" in us'.[35] Similarly, like the New Critics, Barthes is not only resistant to critical methods that seek to ground meaning in the figure of the author, but also to those which appeal to the author's 'hypostases': namely, 'society, history, psyche, liberty', entities which are imagined as dwelling 'beneath the work' and which, once 'found', explain it.[36]

The most significant challenge to the New Critical and poststructuralist assaults on authorial intention came in 1982 with Steven Knapp and Walter Benn Michaels' polemical essay 'Against Theory'. Knapp and Michaels argue that meaning is inextricable from intention: however much material marks may resemble familiar textual signifiers, they cannot be understood as meaningful unless they are intentionally inscribed. They invite the reader to imagine encountering what appear to be lines of poetry etched on the beach: if we count these marks 'as nonintentional effects of mechanical processes (erosion, percolation, etc.)' then to treat them as meaningful would be an invalid projection of agency onto merely contingent phenomena.[37] On the basis of this argument, Knapp and Michaels make the bold claim that since there can be no meaning without intention, 'the meaning of a text is simply identical to the author's intended meaning'.[38] Given this, 'theory' – by which they mean 'the attempt to govern interpretations of particular texts by appealing to an account of interpretation in general' – is a misguided enterprise that should be abandoned. Though few scholars have been willing to accept Knapp and Michaels' arguments wholesale, it is nonetheless apparent in retrospect that, as Reed Way Dasenbrock remarks, '"Against Theory" and the controversy it generated helped usher in the "post-theoretical" era we now seem to be in'[39] – 'post-theoretical' to the extent, at least, that debates over the appropriate hermeneutic or interpretive protocols for textual analysis no longer have the centrality in critical practice that they once had. In their wake, the prevailing tendency has been away from a text-centred focus on the disentanglement of meaning and towards a re-embedding of those meanings within the kind of extra-textual fields that the New Criticism and the 'high theory' of poststructuralism both, in their different ways, sought to bracket out. A return to history, materiality, referentiality, the experiential, the bodily and the real is evident across an array of recently emergent or re-invigorated critical movements, ranging from new historicism and cultural materialism to Marxism, postcolonialism, feminism, queer theory, trauma studies and ecocriticism. Michaels, in particular, has made significant contributions to this contextual or historicist turn in literary studies (most notably his major new historicist work *The Gold Standard and the Logic of Naturalism* [1987]), but if the 'After Theory' controversy played a part in paving the way for this shift, it did so more through its challenge to the dominance of a critical paradigm whose interests tended to exclude extra-textual concerns than through any positive endorsement of those concerns themselves; indeed, Knapp and Michaels insist that they make no claims at all 'about what should count as evidence for determining the content of any particular intention' (intention for them, of course, being synonymous with meaning).[40] Unexpectedly, the field of embodied cognition – and, more distantly, the clockwork writing automata that so suggestively anticipate its interests – indicate ways in which the argument of 'After Theory' might be extended and modified so as to establish a compelling ontological legitimation for the expanded horizons of recent critical study.

The grounds for this legitimation begin to become clear when one considers how a successor of one of Richards' students might respond if asked to write a 'protocol' on a suitably anonymized text by the Cybernetic Poet or BRUTUS. The reader would no doubt be able to give some account of the basic, literal sense of the piece, and might also succeed in tracing some credible patterns of imagery or paths of thematic development, but, once informed of its origins, they would be likely to feel that the exercise had been in some way profoundly futile. As P.D. Juhl observes, there is 'something odd about *interpreting*' a 'computer poem'.[41] Accordingly, McHale describes the 'resentment' that 'anyone who has introduced [interactive, machine-mediated, or machine-generated] poetry to students knows'.[42] Insofar as the function of such avant-gardist strategies is precisely to challenge the reduction of literary reception to a pure matter of determining meaning, McHale's implicit impatience with his students is understandable enough, but, equally, if the urge to decipher familiar, apparently intelligible signs is not simply a convention of certain forms of literary training but an integral element of our very species-being, then the students' resentment is equally excusable. Indeed, to return to the hypothetical example of an exercise in practical criticism being performed on a machine-generated text, the reader's response – which is in this case wholly predicated on the establishment of meaning – would not only feel futile, but *would* be futile, since it would consist of a mere encounter with the 'nonintentional effects of mechanical processes',[43] from which it is as perverse to read off meaning as it is instinctive to do so.

The Cybernetic Poet, BRUTUS and other highly 'delegated' systems of machine writing truly are hypostases of Barthes' dead author: thoughtless, affectless, intentionless beings whose arbitrary manipulations of 'tissue[s] of quotations drawn from the innumerable centres of culture' function, as if by magic, to drain these textual fragments of their significatory power.[44] All that remains for their readers is the possibility of a delirious *dérive* across the smooth surface of the text, in pursuit not of interpretation or decipherment,[45] but of the sheer overwhelming *jouissance* evoked by the material signifier in its all geometric splendour. Despite the best efforts of Barthes and others – including, most notably, Susan Sontag and Gilles Deleuze and Félix Guattari[46] – the notion of a genuinely non-interpretive aesthetics remains less a critical programme, however, than an intriguing thought experiment, one which has in fact only served to demonstrate the inherently interpretive character of every critical statement.

Meaning, then, is the uncircumventable object of reading, and meaning, as we have seen, can be guaranteed only by the agency of an intentional being. Of course, the Cybernetic Poet's 'Soul', BRUTUS's 'Betrayal', or Jackson Mac Low's 'Call Me Ishmael' were not, as in Knapp and Michaels' example, engraved on the beach by some cosmically improbable accident. They each originated, instead, in the actions of a programmer or designer, who presumably had some understanding of the rules and symbols he or she selected for mechanical processing, and some aspiration that, once initiated, the programme would combine these materials in such a way as to produce textual outputs intelligible to a human reader. As Kathleen L. Komar remarks with regard to the Cybernetic Poet, 'the initial reading experience' of Kurzweil and his programming colleagues 'informs the programs they write to create new texts that will produce a similar experience for the reader'.[47] A marginal degree of intentionality, and thus of meaning, can be recuperated in these instances, therefore, but only by appealing to the human agent or agents without whose initiating role no such texts would exist. Why is it, then, that writings by John Donne, Edna St Vincent Millay, or the Reverend G.A. Studdert Kennedy, which I.A. Richards invited his Cambridge classes to respond to in the 1920s, enjoy an intentional and semantic plenitude inevitably withheld from the algorithmically generated text, or available only to the extent that it is guaranteed by the activity of a human programmer? The answer – as the clockwork automata of the eighteenth and nineteenth centuries long ago hinted, and as artificial intelligence has recently demonstrated – is that intentionality can arise only from embodied existence in the referential realm of material objects and relations.[48] Wordsworth thus thematizes the very ontological conditions of possibility of his own poetry when, in 'Tintern Abbey', he strives to summon a vision of his younger self driven by presymbolic, animalistic impulsions to range across the as yet undifferentiated world of phenomena:

> changed, no doubt, from what I was, when first
> I came among these hills; when like a roe
> I bounded o'er the mountains, by the sides
> Of the deep rivers, and the lonely streams,
> Wherever nature led;
> ....
> For nature then
> (The coarser pleasures of my boyish days,
> And their glad animal movements all gone by,)
> To me was all in all. – I cannot paint
> What then I was. The sounding cataract
> Haunted me like a passion: the tall rock,
> The mountain, and the deep and gloomy wood,
> Their colours and their forms, were then to me
> An appetite: a feeling and a love,
> That had no need of a remoter charm,

By thought supplied, or any interest
Unborrowed from the eye. – That time is past. (67–71; 73–84)

To quote Jacques Lacan, it is only from this organic, infantile union with the 'the entirety of things, . . . the totality of the real' that the speaker's capacity for linguistic reflection on his condition can emerge, inscribing 'on the plane of the real this other plane, which we here call the plane of the symbolic'.[49]

Those modes of literary analysis that attempt to separate literary texts out from the spatio-temporal manifold in which they are situated therefore paradoxically exclude the very phenomena that make meaning, and thus criticism itself, possible. This being so, the incorporation of these phenomena – whether they be, say, a soaring rock formation, a ruined religious building, the scars carved on the landscape by the rhythms of industrialization, or the violent upheavals on the streets of revolutionary Paris – into our reflections on literary meaning becomes less a matter of preference than of necessity. The state of embodied intentionality that subtends literary meaning demands, then, a wider consideration of the world through and in which this state develops; but the overdetermined nature of the subject's worldlihood rules out any endorsement of Knapp and Michaels' claim (in the face of the anti-intentionalism of the New Criticism and poststructuralism) that the 'the meaning of a text is simply identical to the author's intended meaning'.[50] If the embeddedness of the human subject in the material conditions of life on earth permits the emergence of its capacity for meaning-making, then, in a recursive movement, it is the privileged manifestation of this capacity – literature – that most powerfully crystallizes these more-or-less contingent and impersonal conditions into meaningful, symbolic form. Such is the virtually infinite variety of these conditions, however, that this intervention on the part of the writer constitutes the coming into being of a field of semantic potential in which meaning may be almost inexhaustibly sought and found. The literary act is the performative announcement of an intention to mean, not the inscription of a singular intended meaning.

Prolegomenon to a robot literary history

Two days after Geoffrey Jefferson delivered the Lister Oration quoted at the beginning of this essay, in which he cast doubt on the likelihood of a machine ever genuinely replicating the human composition of a sonnet, his colleague at the University of Manchester, Alan Turing, was quoted in *The Times* as saying,

> I do not see why it [a computer at the University] should not enter any one of the fields normally covered by the human intellect, and eventually compete on equal terms. I do not think you can even draw the line about sonnets though the comparison is perhaps a little bit unfair because a sonnet written by a machine will be better appreciated by another machine.[51]

By the late 1980s, rudimentary computer-generated poetry was well established. In an essay on a notable early programme, RACTER, Christian Bök speculates that 'the poets of tomorrow are likely to resemble programmers, exalted, not because they can write great poems, but because they can build a small drone out of words to write great poems for us'. He continues: 'What have we to lose by writing poetry for a robotic culture that must inevitably succeed our own? . . . We may have to consider this heretofore unimagined, but nevertheless prohibited, option: writing poetry for inhuman readers, who do not yet exist, because such aliens, clones, or robots have not yet evolved to read it'.[52] Casting an eye towards this far future, Michael L. Johnson wonders, 'what forms beyond the human could evolve, what new kinds of difficult beauty? . . . The rise of silicon "life", silicon intelligence: a Promethean act of technology and language. One may imagine silicon entities floating through deep space, manipulating signifiers beyond human ken'.[53]

The perspective posited in these quotations is articulated at length by Manuel De Landa in his extraordinary book *War in the Age of Intelligent Machines* (1991), which invites us to imagine a future class of

> specialized 'robot historians' committed to tracing the various technological lineages that gave rise to their species. And we could further imagine that such a robot historian would write a different kind of history than would its human counterpart. [. . .] The robot historian . . . would hardly be bothered by the fact that it was a human who put the first motor together: for the role of humans would be seen as little more than that of industrious insects pollinating an independent species of machine-flowers that simply did not possess its own reproductive organs during a segment of its evolution.[54]

As the remarks by Turing, Bök and Johnson suggest, a robot literary history would likely see human beings and their aesthetic interests as similarly marginal to its narrative. What seems increasingly clear, however, is that information processing machines will only conceivably develop the sentience necessary for a literary culture of their own by escaping the prison of nonintentionality; and their only possibility of achieving this is by emulating humans to the extent, at least, of ceasing to dwell in grey boxes on laboratory desks, and emerging, instead, as embodied creatures free to explore the world they inhabit. Any such literary history would no doubt reserve privileged chapters for the clockwork writers of the late eighteenth and early nineteenth centuries, whose embodied forms so strikingly anticipate those of their robotic descendents, as well as the literature generators of our own present, which demonstrate the limits of an existing paradigm, and the necessity of new departures. Both moments also (although, of course, from our robot historian's perspective merely incidentally) cast new light on some central questions in what we will have to learn to call human literary history.

Notes

I am grateful to Melanie Waters and members of the audience at the 'Minds, Bodies, Machines' conference hosted by Birkbeck, University of London in July 2007 for their valuable comments on earlier versions of this essay. I also wish to thank the Franklin Institute in Philadelphia and the State Library of Victoria for permission to reproduce images of two of the automata discussed above.

1. Geoffrey Jefferson, 'The Mind of Mechanical Man', *British Medical Journal* (25 June 1949), 1110.
2. Simon During, *Modern Enchantments: The Cultural Power of Secular Magic* (Cambridge, MA: Harvard University Press, 2002), 121; Jessica Riskin, 'Eighteenth-Century Wetware', *Representations*, 83 (2003), 118.
3. Riskin, 'Eighteenth-Century Wetware', 117–18.
4. Riskin, 'Eighteenth-Century Wetware', 104, 117.
5. E.T.A. Hoffmann, 'Automata', in E.F. Bleiler (ed.), Major Alexander Ewing (trans.), *The Best Tales of Hoffmann* (Mineola: Dover, 1967), 91.
6. John Tresch, 'Fantastic Automata and Their Uncanny Kin in France, 1815–1851', Annenburg Colloquium in European History, University of Pennsylvania, 22 February, 2007. On audience responses to automata in the period, see also During, *Modern Enchantments*, 121–23.
7. Simon Schaffer, 'Enlightened Automata', in William Clark, Jan Golinski and Simon Schaffer (eds), *The Sciences in Enlightened Europe* (Chicago: University of Chicago Press, 1999), 39.
8. Jessica Riskin, 'The Defecating Duck, or, the Ambiguous Origins of Artificial Life', *Critical Inquiry*, 29.4 (2003), 611; and *Science in the Age of Sensibility: The Sentimental Empiricists of the French Enlightenment* (Chicago: University of Chicago Press, 2002), 1–4.
9. For an overview of such continuities, see Chapter One of Aidan Day, *Romanticism* (London: Routledge, 1996), 7–78.

10. Andrew Bennett, *The Author* (London: Routledge, 2005), 54, 57.
11. Christopher Keep, 'Of Writing Machines and Scholar-Gypsies', *English Studies in Canada*, 29.1–2 (2003), 56. See also Bruce Mazlish who, alluding to a myth central to the Romantic movement, remarks that the Jaquet-Droz writer and similar automata extended the promise of 'creative, Promethean force' ('The Man-Machine and Artificial Intelligence', in Ronald Chrisley (ed.), *Artificial Intelligence: Critical Concepts* [London: Routledge, 2000], 138).
12. Vivian Sobchack, '"Susie Scribbles": On Technology, *Technë*, and Writing Incarnate', *Carnal Thoughts: Embodiment and Moving Image Culture* (Berkeley: University of California Press, 2004), 130, emphases in original.
13. Riskin, 'Eighteenth-Century Wetware', 115–17.
14. Gaby Wood, *Edison's Eve: A Magical History of the Quest for Mechanical Life* (New York: Knopf, 2002), xix–xx.
15. References to 'Tintern Abbey' are taken from Stephen Gill (ed.), *The Oxford Authors: William Wordsworth* (Oxford: Oxford University Press, 1984), 131–35.
16. *The Prelude* (1805; published 1850) features a nightmare vision of a London fair, in which Wordsworth puns on the name of John Merlin, the proprietor of a mechanical museum in the West End (from Stephen Gill [ed.], *The Oxford Authors: William Wordsworth* [Oxford: Oxford University Press, 1984], 375–590). See Schaffer, 'Enlightened Automata', 137:

    > All moveables of wonder from all parts,
    > Are here,
    >
    > ....
    > The Bust that speaks, and moves its goggle eyes,
    > The Wax-work, Clock-work, all the marvellous craft
    > Of modern Merlins, wild Beasts, Puppet-shows,
    > All out-o'-th'-way, far-fetched, perverted things,
    > All freaks of Nature, all Promethean thoughts
    > Of man; his dulness, madness, and their feats,
    > All jumbled up together to make up
    > This Parliament of Monsters. (VII.680–81; 685–92)

17. Charles Penniman, 'Maillardet's Automaton', *The Franklin Institute*, at: www.fi.edu/learn/scitech/automaton/automaton.php?cts=instrumentation, 30 March, 2009.
18. Riskin, 'Eighteenth-Century Wetware', 116.
19. Riskin, 'Eighteenth-Century Wetware', 116.
20. Susan Blackmore, *Consciousness: An Introduction* (London: Hodder and Stoughton, 2003), 189.
21. Lewis A. Loren and Eric Dietrich, 'Merleau-Ponty, Embodied Cognition and the Problem of Intentionality', *Cybernetics and Systems: An International Journal*, 28 (1997), 355–56.
22. Notable exceptions are Kathleen L. Komar on Kurzweil's poet ('Candide in Cyberspace: Electronic Texts and the Future of Comparative Literature', *Comparative Literature*, 59.3 [2007], xiii–xiv); Norbert Bachleitner, also on Kurzweil's poet, as well as several comparable projects ('The Virtual Muse: Forms and Theory of Digital Poetry', in Eva Müller-Zettelmann and Margarete Rubik, eds, *Theory into Poetry: New Approaches to the Lyric*

[Amsterdam: Rodopi, 2005], 316–28); and Christian Bök and Josef Ernst on an earlier, more rudimentary programme, William Chamberlain and Thomas Etter's RACTER (Bök, 'The Piecemeal Bard is Deconstructed: Notes Toward a Potential Robopoetics', *Object*, 10 [2002]; Ernst, 'Computer Poetry: An Act of Disinterested Communication', *New Literary History*, 23.2 [1992]). For a useful overview of current critical interests in the field of electronic literature, see N. Katherine Hayles, *Electronic Literature: New Horizons for the Literary* (Notre Dame: University of Notre Dame Press, 2008).

23. Ray Kurzweil, *The Age of Spiritual Machines: When Computers Exceed Human Intelligence* (New York: Viking, 1999), 163.

24. Selmer Bringsjord and David A. Ferrucci, *Artificial Intelligence and Literary Creativity: Inside the mind of BRUTUS, A Storytelling Machine* (Mahwah: Lawrence Erlbaum, 2000), 199–200, emphasis in original.

25. Brian McHale, 'Poetry as Prosthesis', *Poetics Today*, 21.1 (2000), 3.

26. McHale, 'Poetry as Prosthesis', 20.

27. McHale, 'Poetry as Prosthesis', 27–28.

28. McHale, 'Poetry as Prosthesis', 21–22.

29. Bringsjord and Ferrucci, *Artificial Intelligence and Literary Creativity*, xii.

30. Komar, 'Candide in Cyberspace', xiv.

31. Brian McHale proposes 'a kind of Turing Test of poetry: If the machine imitates a human author's linguistic behaviour so perfectly that it can be mistaken for an author, then does it count as an author? And if not, why not?' ('Poetry as Prosthesis', 25 n. 22). I hope to offer a convincing answer to the latter question.

32. Ray Kurzweil, 'A Kind of Turing Test', at: www.kurzweilcyberart.com/poetry/rkcp_akindofturingtest.php3, 30 March, 2009.

33. Carl Sommers, 'By the Way; Inspiration or Computation', *New York Times* (28 November, 1999), at: www.nytimes.com/1999/11/28/nyregion/by-the-way-inspiration-or-computation.html?n=Top%2FReference%2FTimes%20Topics%2FSubjects%2FW%2FWriting%20and%20Writers, 2 April, 2009.

34. William K. Wimsatt and Monroe C. Beardsley, 'The Intentional Fallacy', in William K. Wimsatt (ed.), *The Verbal Icon: Studies in the Meaning of Poetry* (Lexington: University of Kentucky Press, 1954), 3.

35. Roland Barthes, 'The Death of the Author', in Stephen Heath (ed. and trans.), *Image-Music-Text* (London: Fontana, 1977), 143, emphasis in original.

36. Barthes, 'The Death of the Author', 147. The best study of this tradition of attacks on authorial intention remains Seán Burke's classic *The Death and Return of the Author: Criticism and Subjectivity in Barthes, Foucault, and Derrida* (Edinburgh: Edinburgh University Press, 1992).

37. Steven Knapp and Walter Benn Michaels, 'Against Theory', *Critical Inquiry*, 8.4 (1982), 728. The poem is Wordsworth's lyric 'A Slumber Did My Spirit Seal' from the 1800 edition of *Lyrical Ballads*. Knapp and Michaels admit that their 'wave poem' example may seem 'farfetched' but suggest that there '*are* cases where the question of intentional agency might be an important and difficult one', such as the debate over whether computers can 'speak' (728–29, emphasis in original).

38. Knapp and Michaels, 'Against Theory', 724.

39. Reed Way Dasenbrock, *Truth and Consequences: Intentions, Conventions and the New Thematics* (University Park: Pennsylvania State University Press, 2001), 156. On the controversy generated by 'Against Theory' see W.J.T. Mitchell,

Against Theory: Literary Studies and the New Pragmatism (Chicago: Chicago University Press, 1985).

40. Steven Knapp and Walter Benn Michaels, 'A Reply to Our Critics', *Critical Inquiry*, 9.4 (1983), 796.

41. P.D. Juhl, *Interpretation: An Essay in the Philosophy of Literary Criticism* (Princeton: Princeton University Press, 1980), 84, 85, emphasis in original; quoted in Knapp and Michaels, 'Against Theory', 731.

42. McHale, 'Poetry as Prosthesis', 25.

43. Knapp and Michaels, 'Against Theory', 728.

44. Barthes, 'The Death of the Author', 146.

45. Barthes, 'The Death of the Author', 147.

46. See Susan Sontag, 'Against Interpretation', *Against Interpretation and Other Essays* (1966; New York: Picador, 2001); Gilles Deleuze and Félix Guattari, *Anti-Oedipus: Capitalism and Schizophrenia*, trans. Robert Hurley, Mark Seem and Helen R. Lane (1972; London: Athlone, 1984).

47. Komar, 'Candide in Cyberspace', xiv.

48. Selmer Bringsjord, one of the designers of BRUTUS, acknowledges that in order to move towards real creativity, the programme 'would need not only to think mechanically in the sense of swift calculation (the forte of supercomputers like [the chess-playing] Deep Blue), it would also need to think experientially in the sense of having subjective or phenomenal awareness. For example, a person can think experientially about a trip to Europe as a kid, remember what it was like to be in Paris on a sunny day with an older brother, smash a drive down a fairway, feel a lover's touch, ski on the edge, or need a good night's sleep' (*Singularity Summit at Stanford* [Stanford University, 2006], at: http://sss.stanford.edu/others/selmerbringsjord/, 7 April, 2009).

49. Jacques Lacan, *The Seminar I: Freud's Papers on Technique, 1953–54* (New York: Norton, 1988), 262.

50. Knapp and Michaels, 'Against Theory', 724.

51. Elizabeth Wilson, 'Imaginable Computers: Affects and Intelligence in Alan Turing', in Darren Tofts, Annemarie Jonson and Alessio Cavallaro (eds), *Prefiguring Cyberculture: An Intellectual History* (Cambridge, MA: MIT Press, 2004), 43.

52. Bök, 'The Piecemeal Bard is Deconstructed', 17.

53. Michael L. Johnson, *Mind, Language, Machine: Artificial Intelligence in the Poststructuralist Age* (Basingstoke: Macmillan, 1988), 15.

54. Manuel De Landa, *War in the Age of Intelligent Machines* (New York: Zone, 1991), 2–3.

5
Metaphors and Analogies of Mind and Body in Nineteenth-Century Science and Fiction: George Eliot, Henry James and George Meredith

Marie Banfield

The central importance of metaphor to the study of science was stressed by William James when he remarked that science was 'nothing but the finding of an analogy'.[1] The physical, biological and social sciences, modern psychologists have observed, developed on the basis of 'root' or 'founding' metaphors which often went through historical, progressive development.[2] The concept of 'animal spirits', retained in seventeenth-century psychology, was transformed into 'electrical currents' before becoming 'biochemical solutions'.[3] Metaphors, W.H. Leatherdale suggests, are far from decorative but expand the scope of language, providing new concepts. Often regarded as 'the very stuff of language', metaphors can be considered to be 'the instrument of thought'.[4] Drawing together discrete areas of experience, metaphors bring ideas into new juxtapositions and allow a reformulation of experience.[5] These thoughts on the scientific use of metaphor and analogy offer a reminder of the multiple attractions of metaphor and analogy for the writer of literary fiction, where metaphors have greater affective power and yet are also able to introduce new concepts, extend the range of language and enlarge areas of experience, particularly when crossing domains. The ambivalence of metaphor, often regarded with caution by scientists seeking exact definitions and general laws, can offer to the novelist, in his exploration of individual experience, a means of conveying the rich complexity of human existence, with its paradoxes and contradictions.

In the exploratory chapter which follows, I trace some of the root or founding metaphors and analogies that emerged during the nineteenth century to explain the nature of mind and body, and I consider their translation in fiction of the time. I explore the work of psychologists Alexander Bain, George Henry Lewes, James Sully and William James, and novels by George Eliot, Henry James and George Meredith, writers regarded by their contemporaries as 'psychologists'. Literary discussion focuses chiefly on *Middlemarch* (1872), *The Portrait of a Lady* (1881) and *Diana of the Crossways* (1885), novels which coincidentally all examine the same human dilemma: the making and the breaking of an unhappy marriage.

This group of literary and scientific writers were materially connected as well as ideally related, sharing professional networks and in some cases friendships. Several wrote for the same journals, notably the *Westminster Review* and the *Fortnightly Review*. The professional and personal commitment shared by Eliot and Lewes is well known and needs little comment, except to note that after Lewes's death Eliot edited the final two volumes of *Problems of Life and Mind* (1879), 'generously aided' by Sully.[6] The high regard in which Meredith held the writing of Lewes and Bain can be inferred from his letters.[7] The friendship of Sully with Eliot and with Meredith is described in the psychologist's autobiography.[8] Henry James corresponded with Meredith and visited his Box Hill home.[9] His brother, William James, was a member of the British Society for Psychical Research, founded in 1882. His *Principles of Psychology*, which drew together ideas formed during the 1880s, has been described as a 'landmark contribution' to the philosophy of mind.[10]

An early but memorable description of mind's struggle to find metaphors is found in William Cowper's poem 'The Task', first published in 1785:

> . . . The shifts and turns,
> The expedients and inventions multiform,
> To which the mind resorts, in chase of terms
> Though apt, yet coy, and difficult to win –
> To arrest the fleeting images that fill
> The mirror of the mind, and hold them fast.[11]

Cowper's metaphor of mind as mirror signifies the lucidity, the faithfulness, and the fullness of representation that the poet strives to achieve, but the image fails to convey the mental struggle which is acknowledged to be central to the production of images. Thomas Hobbes, in *Leviathan* (1651), posited a more complex and dynamic model of mind when he envisaged the 'motion, agitation, or alteration which worketh in the brain'.[12] Hobbes, Romanes suggests, was the first person to earn of the title of 'psychologist', having anticipated discoveries in physiology and psychology which took two centuries of scientific research to establish, and which produced the idea of mind as motion.[13] Seeking exact definitions in science, and words 'purged from ambiguity', Hobbes in *Leviathan* rejected metaphors as *'ignes fatui'*.[14]

Lewes, in an early essay on Spinoza, cautioned against confusing metaphor with fact, instancing the image of mind as mirror which, he argues, gives 'no faithful reflection of the world' but only 'a faithful report of its own states':

> The great mistake lies in taking a metaphor for a fact, and arguing as if the mind were a *mirror*. It is no mirror; it gives no faithful reflection of the world; it gives only a faithful report of its own states, as excited *by the world* . [. . .] The mind is not a passive mirror reflecting the nature of things, but the partial creator of its own forms.[15]

In perception, Lewes notes, there are changes only in the percipient. Lewes is aware of the potency of metaphors which, embedded in language, preserve some of their force even when the concepts with which they are associated become outdated:

> Throughout the course of metaphysical speculation, no one hesitated to believe that images of some kind were formed in the eye, as in a mirror, and thence 'transmitted to the mind' . . . How many physiologists and philosophers of the present day really entertain this notion of images being transmitted to the brain, it is difficult to say, because ordinary language is so impregnated with the old conception, that men who have long rejected it as an hypothesis may still use it as a metaphor.[16]

Such metaphors, Lewes observes, maintain their viability even among those who understand their lack of scientific veracity. The image of mind as mirror is rejected by William James in more complex terms:

> The knower is not simply a mirror floating with no foot-hold anywhere, and passively reflecting an order that he comes upon and finds simply existing. The knower is an actor, and co-efficient of the truth on one side, whilst on the other he registers the truth which he helps to create.[17]

The 'knower', James argues, is not a reflector of the world around him but an agent who helps to create the reality he experiences.

Analogy, Leatherdale observes, involves resemblance of relations and likeness with differences.[18] Metaphorical comparisons, Bain notes in

The Senses and the Intellect, can be 'real and substantial', such as the moon's gravitational pull and the fall of a stone; these are not simply illustrative but represent scientific discoveries. Others are of a mixed nature, when one term belongs to 'the province of mind' and the other to physical phenomena. Metaphors of the second kind, Bain observes, can represent what is obscure or concealed from view. Many social phenomena, he notes, are never conceived except in terms of material analogy.[19] The poetic and fictive possibilities of metaphor implicit in Bain's account are revealed when he notes the capacity of metaphor to reveal hidden thoughts and feelings, and when he describes the ability of association to facilitate the retention of narrative detail.[20]

In the eighteenth century, metaphors taken from science and technology dominated the writing of David Hartley, as Rick Rylance describes: 'the action of pendulums, particles in sounding bodies, the transmission of heat and light, . . . musical strings, mechanical friction'. Hartley's mechanical wires, Rylance notes, evolved into 'organic filaments and tissues', finding their direct descendent in George Eliot's descriptions of threads, tissues and vibrations.[21] In *Middlemarch*, the work of French anatomist and physiologist Marie François Xavier Bichat fascinates Dr Lydgate, who is determined to expand the scientific basis of the medical profession. The Frenchman's conception, that living bodies were not just connected organs but webs of tissue, uncovers for Lydgate 'new connexions and hitherto hidden facts of structure' that needed to be taken into account in any medical diagnosis. The change represents for Lydgate a technological revolution, like the 'turning on of gas-light' in a 'dim, oil lit street', a movement which would alter medical theory and practice.[22] Eliot's intricate description of the processes of scientific research reveals, as Michael Davis observes, her 'profound sense of the varied powers of the mind'.[23]

The foundation for a new science of mind was laid by Bain in the 1850s, in *The Senses and the Intellect* and *The Emotions and the Will*, two volumes which united earlier psychological approaches, bringing together the associationism of Hartley and the sensory motor-physiology of experimentalists such as Johannes Müeller.[24] In *The Senses and the Intellect*, Bain rejects the 'old and widely prevalent' view of mind[25] as a receptacle of sense impressions, or a store chamber:

> The organ of the mind is not the brain by itself; it is the brain, nerves, muscles and organs of sense. [. . .] It is, therefore, in the present state of our knowledge, an entire misconception to talk of a *sensorium* within the brain, a *sanctum sanctorum,* or inner chamber,

where impressions are poured in and stored up to be reproduced in a future day. There is no such chamber, no such mode of reception of outward influence.[26]

Mind, Bain argues in 'Common Errors on the Mind', is not a different fact from the body: 'not a feeling can arise, not a thought can pass,' he insists, 'without a set of concurring bodily processes'.[27] According to this modern view, Bain observes, the brain is only responsible in part for nervous action.

The intimate, powerful connection between the physiological body and mental and emotional life, suggested by Bain's psychology, is conveyed in Meredith's novel, *The Ordeal of Richard Feverel*, published in 1859. Here physiological and emotional elements are used as correlates when Richard's mental condition is compared to the physical condition of a drowned man being fished out of the river. The physical pain and strenuous effort endured by a body struggling to revive the life processes in an organism already drying out and surrendering to the forces of decay, serves to signify the mental and emotional agony incurred by Richard as his dormant feelings begin to reassert themselves:

> They say, that when the skill and care of men rescue a drowned wretch from extinction, and warm the flickering spirit into steady flame, such pain it is, the blood forcing its way along the dry channels, and the heavily-ticking nerves, and the sullen heart – the struggle of Life and Death in him – grim Death relaxing his grip: such pain it is, he cries out no thanks to them that pull him by inches from the depths of the dead River.[28]

The material relation between mind and body and its vital forces is stressed in the anatomical detail of the analogy: the pressure of blood in arteries already closing, pushing its way by force to a heart that is slowing, conveys the desperate nature of Richard's mental and emotional condition. The 'ticking' of the nerves suggest the winding down of an internal clock, the rhythmic ebbing away of life. The material, anatomical details situate Richard's mental and emotional struggle within his body.

Metaphors used by Bain to explain and illustrate his concept of mind and body are taken largely from existing sciences: from physics, physiology and mechanics, as well as from the new science of thermodynamics. In his essay 'On the Correlation of Force in its Bearing on Mind', Bain argues that the correlation of mind and body must involve the nerve force.[29] In *The Senses and the Intellect* he describes the arrangement of the nervous system in natural metaphors, as a tree or a river system, but he also illustrates it in modern technological terms as 'the course of a railway train' or a telegraph system in which the system of wires 'might be formed to represent exactly what takes place in the brain'.[30] Consistently, Bain draws analogies between mental force and electricity. The familiar properties of magnetism and electricity are borrowed to explain the unknown relation between body and mind. The telegraphic wire provides a metaphor for the nerve that runs from the brain to any part of the body; and the voltaic battery, with its generation of electric power from corroding acid, illustrates the action of the grey substance of nerve centres, charged with blood.[31] Electricity, Bain points out, offers a more suitable analogy for mental force than Hartley's earlier vibrations of sound but he notes that although it is more subtle it does not represent identity. If it is a kind of electric force, it is, he argues, a 'distinct species', for there is no closed circuit, and the nerve fibre, unlike wire employed in an electric circuit, is consumed in maintaining the current.[32] In *The Senses and the Intellect*, vitality is explained as 'a collocation of the forces of inorganic matter for the purpose of keeping up a living structure'.[33] Bain describes the active force of the brain as a 'current of energy', a 'stream of power', a 'common stream of mental activity'.[34] 'The concurrence of Sensations in one common stream of consciousness,' he observes, allows different senses to be readily associated.[35] Central to Bain's theory of mind is the concept of 'Spontaneous Energy', an inherent, active power, a surplus nervous energy, which has no initial purpose but to expend itself. It is 'the surplus nervous power of the system discharging itself without waiting for the promptings of sensation'.[36]

Analogies drawn from this new energetic model of consciousness are employed by Eliot in *Middlemarch* alongside older images of mind as mirror or receptacle. The mind of Casaubon is seen spatially by Dorothea, in terms of a 'labyrinthine extension' and an 'ungauged reservoir'.[37] She fears her own thought will appear 'a poor two penny mirror' in comparison, her emotions and experiences a 'little pool' to his great lake.[38] The image of water, still and landlocked, suggests a state of stasis, contrasting Bain's idea of the nervous system as a stream continually moving and perpetually restoring itself in a kind of 'equilibrium'.[39] Casaubon husbands his resources, and guards his mental and physical expenditure; believing himself to feed too much on 'inward sources', fearing that he is 'using up' his eyesight.[40] When he attempts to broaden his experience, and give himself to 'the stream of feeling', he finds it a 'shallow rill'.[41] Casaubon seems dislocated; faced with the confusion and speed of modern change, he attempts to reconstruct the world as it was but feels disembodied, like an 'ancient wandering about the world'.[42] He appears to belong to an older cosmology: the centre of his own world, he regards others as his satellites, created providentially for him. Aware of the forces of modern life but removed from them, Casaubon wrestles with the 'universal pressure' which, the narrator predicts, will one day prove too much for him.[43]

In contrast to the still reservoir and 'shallow rill' of Casaubon's mental and emotional life, the mind of Lydgate is associated with movement, complexity and above all with energy. Mentally disciplined, Lydgate's mind controls well the petty day-to-day inconveniences until deteriorating circumstances cause a 'second consciousness', to confront him with his 'degrading preoccupation' and 'wasted energy'.[44] Mind and body in Lydgate work together, in close and sometimes disturbing union; frustration and confusion on occasion urge violence: 'paralysed by opposing impulses . . . he wanted to smash and grind some object on which he could at least produce an impression'.[45] Ultimately his spontaneous energy, or nervous surplus, is expended unproductively. Financial and domestic pressures exhaust him, leaving no 'free energy' for 'spontaneous research and speculative thinking'.[46] His mental and emotional anxiety is registered on his face, his 'energetic nature' turns into 'dogged resistance'.[47] Lydgate's 'energy', we are informed by the narrator, falls short of the task.[48]

If the concept of energy supports the character of Lydgate, it is the specific force of electricity which is seen to animate Dorothea. The analogy aptly stands for the fluidity and quickness of her temperament, for that 'full current of sympathetic motive in which her ideas and impulses were habitually swept along'. Dorothea recognizes the deep connection between her mental life and her physiological nature: she does not want her knowledge to be an adornment but to be worn 'loose from the nerves and blood that fed her action'.[49] When Casaubon visits the house and leaves pamphlets for Dorothea, it is as if an 'electric stream' goes through her.[50] When his letter proposing marriage arrives, it releases currents of feeling which seem to speak of freedom and fulfilment, a direction for her energies:

> Her whole soul was possessed by the fact that a fuller life was opening before her: she was a neophyte about to enter on a higher grade of initiation. She was going to have room for the energies which stirred uneasily under the dimness and pressure of her own ignorance and the petty peremptoriness of the world's habits.[51]

The narration distances itself from Dorothea's response, building upon doubts already seeded in the narrative through contrasting metaphors of stasis and of animation associated with Casaubon and Dorothea, images which signal a fundamental incompatibility. That Eliot appreciates the potential danger and possible treachery of metaphor is noted by Felicia Bonaparte who, in her introduction to *Middlemarch*,[52] reminds us of the narrator's reflection in relation to Casaubon that we all 'get our thoughts entangled in metaphors, and act fatally on the strength of them'.[53]

Metaphors and analogies drawn from physical and physiological science in Eliot's novel become a flexible fictive instrument. New ideas and older concepts are juxtaposed, provoking thought, differentiating characters and ideologies, offering a key to personality and identity while suggesting deeper levels of experience. The use of metaphors drawn from contemporary science recognizes and utilizes a new conceptual understanding of the relationship between mind and body; a dynamic concept of mind based on recognition of the interdependence of mental, emotional and sentient life, and on an understanding of consciousness as a force, a current, a stream of impressions.

The concept of mind as linked to the physical state of the body, fluid, dynamic, and subject to fluctuations of energy, was developed by psychologists during the 1870s. Lewes, in *Problems in Life and Mind* (1879), described each mental state as belonging to a finely integrated system:

> The brain is simply one element in a complex mechanism, each element of which is a component of the Sensorium, or Sentient Ego. We may consider the several elements as forming a plexus of sensibilities... no one of them can be active without involving the activity of all the others.[54]

The inner life is seen as 'variously blended', with external and internal stimulations, and past experiences modifying the effects of present experiences.[55] 'One stroke,' he notes, 'sets the whole vibrating.'[56] Sensibility, Lewes argues, is not a '*res completa*', to be perceived, measured, or laid hold of, a 'vital spirit' or 'electric current', but an energy manifest in phenomena.[57] Objects, Lewes argues, are constituted not only of present feelings but are syntheses of past and present, made into new constructions.[58] Objectivity and subjectivity in Lewes's writing are brought into the closest of relationships. '*All physical facts*,' he observes in 'The Course of Modern Thought', are '*mental facts expressed in objective terms*, and *mental facts are physical facts expressed in subjective terms*.'[59]

The closest of all analogies, Lewes reflects, is the analogy between life and mind.[60]

Each human being, Sully notes in 'The Aesthetics of Human Character', is not only a subjective mind but is also part of the objective world, as much as a tree or a rock. Our inner mental states are revealed to others by means of material investiture.[61] There are 'three great avenues,' he advises in *Sensation and Intuition*, which allow us to look into the minds of others: the 'expressive look', the 'resulting action' and the 'descriptive word'.[62] The dramatist, he observes, has the advantage of representing visible and audible impressions yet the novelist has the freedom to describe invisible, inaudible thoughts and impulses.[63] The audience or reader adopts an objective and a subjective role, acting as both 'spectators' and 'judges'.[64]

William James, in an early essay on introspective psychology, acknowledges the work of English psychologists but speaks of vast tracts of inner life which have been overlooked: unnamed or *'dumb psychic states'*.[65] The mind, James contends in an essay on Herbert Spencer, has laws not only of logic but also of 'fancy, of wit, of taste, decorum, beauty, morals'.[66] The relations of feeling, James observes, are so innumerable that language cannot adequately do justice to their shades and nuances. Enriching and refining Bain's earlier metaphor of mind as a 'stream of consciousness', James describes 'thought's stream' not as discrete, comprised of separate parts, but as 'sensibly continuous, like time's stream'.[67] Nascent images, though not a subject for observation, are among the objects of the stream.[68] The object of any thought, James argues, is its entire deliverance; every word is 'fringed' and each sentence 'bathed' in 'a halo of obscure relations'.[69] This fringing, he notes, may include a feeling of continuity with earlier thoughts. James's theory of stream of consciousness, suffused with all the experiences of the senses, Jill Kress argues, suggests a coherence that washes over 'distinctions or separations'.[70] The ultimate principle of Jamesian consciousness, she reflects, appears to lie in its creative capacity. Mind, as William James acknowledges, becomes 'a theatre of simultaneous possibilities'.[71]

In Henry James's depiction of Isabel Archer in *The Portrait of a Lady*, the intricacy, the richness, the fluency and the intensity of mental experience, as described in William James's psychology, finds fictive expression. Isabel's mental life is active, one with her emotional life, and is read as much through her bodily gestures, movements, expressions and unspoken thoughts as through her spoken words. In regard to herself as well as to those around her, she is, to use Sully's terms, both spectator and judge. Isabel's natural observational skills compensate for a lack of experience. She has, we are told, an eye that 'denoted clear perception'.[72] Her senses are acute and mediated by an active intelligence. Isabel is, as the narrator remarks, a 'young person of many theories';[73] 'full of premises, conclusions, emotions'.[74] Feeling, she assures Ralph, is natural for a sentient being and cannot be distinguished from seeing.[75] In Isabel, a consciousness is presented that is constantly at work. Complex and capable of great contradictions, she is exacting and indulgent, curious and fastidious, vivacious and indifferent. She is a 'delicate, desultory, flame-like spirit' and 'a creature of conditions'.[76] She is the product of her outer material circumstances but also of immaterial inner strivings. If she fails, Philip Sicker observes, it is not because she lacks awareness of 'perpetual change' within herself, but because she remains blind to the complexity of the minds of others.[77]

The question of the relation between inner and outer life, between material circumstance and self is explored directly in the novel, in conversation between Isabel and Madame Merle. For the older woman, material circumstances form a shell around the individual that must be taken into account.[78] Each individual is made up of a 'cluster of appurtenances', of which clothes and possessions constitute a major part, an expression of what we call self. Isabel argues against such a reduction of life, noting that nothing that belongs to her is to be her measure, but is to be regarded simply as 'a barrier'.[79] If to live 'in the Jamesian sense', as Sicker observes, is to choose and to reject experiences through which social identity is projected, then Isabel lives through the acts of discrimination and choice.[80]

In contrast to Madame Merle's aridity, Isabel's mental and emotional life is shown to be rich, febrile and fluent. Past and present impressions form a single flow as things forgotten are brought back to mind and present concerns drop out of sight.[81] Her perceptions are acute and, she fears, on occasion are so innumerable that her conscious judgment cannot cope.[82] The alertness of her intellectual life is matched by the vibrancy of her sensational and emotional experience. Vibration, we are told, is 'easy to her'. Following her dismissal of Caspar Goodwood, she trembles like a 'smitten harp'.[83] Her conscious emotion is one of relief but beneath this is perceived an opposing feeling: exultation at the exercise of power. Beyond Isabel's intellectual theorizing are glimpsed areas of experience outside of the range of conscious control and articulation, the vast tracts of inner life described by William James.

Against Isabel's intuitive, thoughtful appraisal is placed the relentless journalistic scrutiny of Henrietta Stackpole. She emerges, from the perceptions of Ralph Touchett, as one unaware of the complexities and anomalies of life, a woman for whom human beings remain 'simple and homogeneous organisms'.[84] Her eyes, Ralph notes, seem not to give access to her inner life but to be large reflecting surfaces: like 'large polished buttons . . . of some tense receptacle', in which could be seen reflected the surrounding objects.[85] In his analogy older metaphors are revived. Describing the exteriority of Henrietta's vision, these images suggest a mind functioning chiefly as a reflecting glass, and partly as a container or store for impressions, in contrast to the complex, more reflexive mind of Isabel.

In contrast with the fluidity of Isabel's mind and personality, the two men brought into closest relationship with her are portrayed in rigidly material terms. Caspar Goodwood is, for Isabel, the 'stubbornest fact' she knows.[86] His inflexibility of purpose is seen in his physique; his body, 'too straight and stiff', his jaw 'too square and set'. There is in him, Isabel perceives, a 'want of easy consonance with the deeper rhythms of life'. Associating him with metal, hard, inflexible and reflective, she sees Goodwood pieced together as a suit of armour, his eyes impenetrable, visible only through a visor.[87] The eyes of Osmond are intelligent, the eyes of an observer and a dreamer, but they too are hard. Osmond is of a finer grain but he also is metallic: a 'fine gold coin'; a medal struck off for a special occasion.[88] Isabel, he regards as silver plate; his desire is to 'tap her imagination with his knuckle and make it ring'.[89]

The play of consciousness within Henry James's novel is far removed from the concept of mind as a passive reflecting surface. We perceive in Isabel Archer the natural and unconstrained autonomy which William James attributes to the mind, which from 'birth upward' has both 'spontaneity' and 'a vote', is 'in the game' and 'not a mere looker-on'. She represents his 'knower' who is 'not simply a mirror' but an 'actor' who helps to create the truth she registers.[90] William James's concept of mind as a stream of consciousness offers a powerful analogy which cuts deeply into Henry James's fiction; its influence is felt in the fluidity of its characterization and the complexity of the emotional experience it describes.

Images of an earlier association psychology, of threads, links and chains, disputed by William James,[91] are retained in Sully's psychology to represent human consciousness. Sully, in *Outlines of Psychology*, describes mind as 'the inner smaller world', the 'mikrokosm' within the 'makrokosm' of the external, larger world.[92] Knowledge, he argues, begins with the senses, and develops as life advances, increasing in complexity with the exercise and the development of the brain.[93] States of mind are described as a mass of mental phenomena, an intricate chain of mental operations. At any one time the mind is 'a tangle of psychical states or threads of psychical processes',[94] the unravelling of which is the task of the psychologist, who has available to him two approaches: the first is direct, internal and subjective; the second is indirect, external and objective. The first is possible through a process of introspection; the second relies upon close observation of material phenomena.[95] Observation, Sully notes, requires a restraining of the imagination, a setting aside of personal prepossessions in order to direct the mind upon the object of study, as it is presented to the senses.[96]

Lewes, like Bain and William James, describes the mental process as a 'stream', but as a flow of 'sentience' rather than of consciousness, a system in which desires, emotions and instincts make up the greater part of the flow.[97] Though consciousness is the salient state, Lewes stresses the importance of unconscious forces which for the biological psychologist play the larger part in the life of the mind.[98] Beyond conscious thought, outside the range of introspection, he locates many fugitive, '*masked*' or 'latent' states;[99] the 'silent processes of growth' which make up character.[100] The mystery of consciousness, of feeling and existence, he notes is signified in the fact that we are able to name them only in metaphor.[101]

George Meredith, in exploring the inner life of his protagonist in *Diana of the Crossways,* adopts a double approach analogous to methods described in contemporary psychology. Diana Warwick is a writer who studies the external world through practised observation while she examines her own mind through introspection. This flexible approach becomes in the novel a powerful fictive tool: a means of reading interior life while detailing the exterior world. Diana, as Meredith explained to R.L. Stevenson in March 1884, was 'partly modelled' on Caroline Norton, who was a novelist and poet and early feminist writer.[102] The main events of the narrative parallel Norton's life, the details of which give a moving account of a woman's suffering under English law in the early nineteenth century. Yet, the emphasis is less on public life than on 'internal history', which the narrator insists is 'the brainstuff of fiction'.[103] Capturing the fluidity of mental and emotional states, the altering moods and shifting intellectual boundaries, the novel explores the sentient life of the mind as it probes beyond boundaries of conscious thought.

As novelist, Diana is aware of the intricacy of the processes of the human mind: she sees 'living men and women' as 'too various in the mixture fashioning them', to be rendered convincingly in a formal sketch. The mind, constantly moving, she insists, cannot deal with 'protracted description':

> I may tell you his eyes are pale blue, his features regular, his hair silky, brownish, his legs long, his head rather stooping (only the head), and his mouth commonly closed; these are the facts, and you have seen much the same in a nursery doll.[104]

Although she understands well her creative method, her aesthetic is more ambivalent: she criticizes the world for producing a 'miry form of art' which tries to read nature's depths in its 'muddy shallows', yet she shudders at her own power to 'draw from dark stores'.[105] A parenthetical remark, '(we write on darkness)', conveys her awareness of the unconscious depths upon which a writer has to draw.[106]

Her mental and emotional crisis after her flight to 'The Crossways', following the collapse of her marriage, is shown to be a struggle of unconscious drives as well as conscious emotions: her sensations direct 'her rebellious wits' and her thoughts distil 'much poison'.[107] Her altering mood, the expansion and digressions of her thought, the physical details of her fatigue, and the fact that the scene takes place at night, all give substance to a complex psychological portrait. Diana's mental and emotional state is made visible chiefly through the images her mind yields. The chapter begins and ends with an image of her mind as 'a steam-wheel', a metaphor which signifies both a process of relentless, mechanical activity and the work and productivity of the brain.

In analysing her situation of entrapment, Diana makes comparisons with other states of captivity and punishment. Almost in a delirium, her imagination produces a series of nightmarish visions as the rational part of her reverie is overtaken. Threads of association intersect one another, forming from disparate experiences an intricate web or 'tangle of psychical states'.[108] She glimpses her husband behind a mask and sees herself as a strange 'half-known, half-suspected, developing creature' that claimed to be Diana but was unlike the woman 'deformed by marriage'.[109] Discordant images converge upon the hypothesis that if she had known love she would not now be as a 'caged-beast'.[110] Metaphors are not elaborately constructed but seized on by the mind in synecdochal form. When the civilized world can offer no adequate analogy for her experience, a detail from the Roman circus fills the gap:

> She was in the arena of the savage claws, flung there by the man who of all others should have protected her from them.[111]

The mental process, like Lewes's stream of sentience, is directed chiefly by instinct and feeling. The resolution that emerges in the impressionable half light of daybreak, not to go into exile but to become 'the first martyr of the modern woman's cause', is emotionally rather than logically driven.[112] Her decision is taken in spite of a contrary warning from an inner voice, which she takes to be reason.[113] The process of self scrutiny is shown to be rigorous and painful but ultimately fruitful:

> By closely reading herself, whom she scourged to excess that she might in justice be comforted, she gathered an increasing knowledge of our human constitution, and stored matter for the brain.[114]

The experience gained in her suffering deepens her self-knowledge and her understanding of human nature and increases her mental strength.

A mixture of conscious and unconscious motives directs the pivotal act of the narrative, when Diana sells a state secret to the Press. In retrospect, the action seems to her inexplicable. She attempts to explain it through the metaphor of a loaded gun:

> When I drove down that night to Mr. Tonans, I am certain I had my clear wits, but I felt like a bolt. I saw things, but at too swift a rate for the conscience of them . . . I was a shot out of a gun.[115]

A state of heightened consciousness is described in which things are seen clearly but in which impressions pass too rapidly through the mind for measured reflection or moral discrimination. The analogy of a charged gun is used by Lewes in *Problems of Life and Mind* to describe the formation of acts of sudden impulse:

> Certain tendencies are organised, certain preparations are disposed, which only await some incidental stimulus to start into conceptions and acts which seem incongruous and sudden to the man himself as to the onlooker. The gun has been silently charged, and at a touch of the trigger it explodes.[116]

A similar image seized upon by Diana suggests a compulsive act beyond the control of conscious thought, dependent on motives deeply rooted in her mind, beyond the scope of rational explanation.

Both narrator and protagonist in *Diana of the Crossways*, sophisticated readers, writers and analysers of language, are consciously aware of the power of metaphor, perceiving the ways in which it can function within individual relationships as well as in broader social and cultural interactions. In her relationship with Dacier, Diana senses a fundamental discrepancy between their ways of regarding the world, revealed at perceptual and linguistic levels. His 'grotesque' and 'anti-poetic' metaphors appear isolating, in a manner she finds chilling.[117] Her own metaphors provide her with a sanctuary, a space in which to reflect on life in all its relations:

> Metaphors were her refuge. Metaphorically she could allow her mind to distinguish the struggle she was undergoing, sinking under it. The banished of Eden had to put on metaphors, and the common use of them has helped largely to civilize us. The sluggish in intellect detest them, but our civilization is not much indebted to that major faction. Especially are they needed by the pedestalled woman in her conflict with the natural.[118]

Metaphors for Diana have a social and cultural function; they help to tame and to advance the world. Women, in particular, are felt to have need of them, not only to explain life but to make it more bearable. This discussion, as Gillian Beer observes, contains the idea that metaphor civilizes because it helps to uncover levels of experience that would otherwise remain 'inchoate and beyond the reach of reason'.[119] In Bain's terms, it makes the hidden visible.

The close connection made in Meredith's novel between material reality and metaphor is considered by Kurt Danziger within a scientific context when he suggests that when Bain uses metaphors of 'tracks' and 'steam', he is not engaging in picturesque language but believes the mind to be a kind of 'tracked energy'.[120] A founding metaphor of modern psychology, the validity of this metaphor was readily accepted not only within the scientific world but also outside it. The novelists discussed in this chapter, although they continued to draw on earlier, static images for mind, using metaphors which remained linguistically viable and fictively useful, increasingly turned to more dynamic images to describe mental processes. Ideas of 'collocation of forces', 'current of impressions', 'stream of consciousness', 'stream of sentience' provided analogies for mental and emotional states that proved irresistible to writers seeking more accurate ways to interpret human mind and action. These metaphors and analogies provided a tool for the nineteenth century novelist before the concept was consciously adopted and developed by twentieth-century modernist writers who, as Rylance persuasively argues, built their aesthetic directly on the psychology of William James, erasing earlier traces.[121]

Notes

1. David E. Leary, *Metaphors in the History of Psychology* (Cambridge: Cambridge University Press, 1994), 19.
2. Leary, *Metaphors in the History of Psychology*, 14–15.
3. Leary, *Metaphors in the History of Psychology*, 16.
4. W.H. Leatherdale, *The Role of Analogy, Model and Metaphor in Science* (Amsterdam: North Holland Publishing, 1974), 103.
5. Leatherdale, *The Role of Analogy, Model and Metaphor in Science*, 93.
6. See 'Prefatory Notice' to *Problems of Life and Mind: Mind as a Function of the Organism*, by George Lewes (London: Trübner, 1879). After Lewes's death Eliot arranged his remaining notes for *The Study of the Psychology* and *Mind as a Function of the Organism*, which were published in 1879 (Rosemary Ashton, *G. H. Lewes: An Unconventional Victorian* [London: Pimlico, 2000], 279).
7. See C.L. Cline (ed.), *The Letters of George Meredith* (Oxford: Clarendon, 1970), 2:578–9. In a letter written to Miss Ida Benecke on 23 September, 1879, Meredith recommends the reading of Lewes and 'more particularly Bain', in preparation for some lectures on mental philosophy she is to attend, suggesting his familiarity with and appreciation of their work.
8. Sketches of George Meredith and George Eliot are included in 'Portraits of Friends', the second part of Sully's autobiography. See James Sully, *My Life and Friends: A Psychologist's Memories* (London: Fisher Unwin, 1918).
9. In a letter to James written on 11 May 1895, Meredith provides him with elaborate directions from London to Flint Cottage, Meredith's home near Box Hill. As an extra allurement, Meredith gives details of his wine cellar (See Cline, *The Letters of George Meredith*, 3:1192). On 4 June, 1897, Meredith reassures James that there is always a bed for him at Flint Cottage (Cline, *The Letters of George Meredith*, 3:1267).
10. Gerald E. Myers, *William James: His Life and Thought* (New Haven: Yale University Press, 1986), 2. This chapter focuses on the earlier essays of William James and on his 'Psychology: A Briefer Course', a condensed and revised version of *Principles of Psychology*, all of which are found in *William James: Writings 1878–1899*.
11. William Cowper, *The Task: A Poem* (London: James Nisbet, 1878), bk 2, 61.
12. George John Romanes, *Mind and Motion and Monism* (London: Longmans, Green, 1895), 2.
13. Romanes, *Mind and Motion and Monism*, 1.
14. Thomas Hobbes, *The Leviathan* (Toronto: Broadview Literary Texts, 2002), 38–39.
15. George Henry Lewes, 'Spinoza's Life and Works', *Westminster Review*, 39 (1843), 400–1.
16. George Henry Lewes, *The Physiology of Common Life* (Leipzig: Bernard Tauchnitz, 1860), 2:229.
17. William James, 'Spencer's Definition of Mind', in Gerald E. Myers (ed.), *Writings, 1878–1899* (New York: Library of America, 1992), 908.

18. Leatherdale, *The Role of Analogy, Model and Metaphor in Science*, 2.
19. Alexander Bain, *The Senses and the Intellect* (London: John W. Parker, 1855), 530–31.
20. Bain, *The Senses and the Intellect*, 543.
21. Rick Rylance, *Victorian Psychology and British Culture, 1850–1880* (Oxford: Oxford University Press, 2000), 85–86.
22. George Eliot, *Middlemarch* (Oxford: Oxford University Press, 2008), 139.
23. Michael Davis, *George Eliot and Nineteenth-Century Psychology: Exploring the Unmapped Country* (Aldershot: Ashgate, 2006), 1.
24. Robert M. Young, *Mind, Brain and Adaptation in the Nineteenth Century* (Oxford: Oxford University Press, 1990), 5–6.
25. Bain, *The Senses and the Intellect*, 332.
26. Bain, *The Senses and the Intellect*, 61.
27. Alexander Bain, 'Common Errors on the Mind', *Fortnightly Review*, New Series, 4 (1868), 160.
28. George Meredith, *The Ordeal of Richard Feverel* (London: Penguin, 1998), 2:243–44.
29. Alexander Bain, 'On the Correlation of Force in its Bearing on Mind', *Macmillan's Magazine*, 16 (1867), 373.
30. Bain, *The Senses and the Intellect*, 30.
31. Bain, *The Senses and the Intellect*, 57.
32. Alexander Bain, 'The Feelings and the Will Viewed Physiologically', *Fortnightly Review*, 3 (1866), 578.
33. Bain, *The Senses and the Intellect*, 60.
34. Bain, *The Senses and the Intellect*, 79, 325.
35. Bain, *The Senses and the Intellect*, 359.
36. Bain, 'The Feelings and the Will Viewed Physiologically', 587.
37. Eliot, *Middlemarch*, 22.
38. Eliot, *Middlemarch*, 23.
39. Bain, 'The Feelings and the Will Viewed Physiologically', 580.
40. Eliot, *Middlemarch*, 16.
41. Eliot, *Middlemarch*, 58.
42. Eliot, *Middlemarch*, 16.
43. Eliot, *Middlemarch*, 78.
44. Eliot, *Middlemarch*, 608–9.
45. Eliot, *Middlemarch*, 621.
46. Eliot, *Middlemarch*, 628.
47. Eliot, *Middlemarch*, 697.
48. Eliot, *Middlemarch*, 714.
49. Eliot, *Middlemarch*, 80.
50. Eliot, *Middlemarch*, 35.
51. Eliot, *Middlemarch*, 41.
52. Felicia Bonaparte, introduction to *Middlemarch*, by George Eliot (Oxford: Oxford University Press, 2008), xxxvi.
53. Eliot, *Middlemarch*, 79.
54. George Lewes, *Problems in Life and Mind: Mind as a Function of the Organism* (London: Trübner, 1879), 77.
55. Lewes, *Problems in Life and Mind: Mind as a Function of the Organism*, 101.
56. Lewes, *Problems in Life and Mind: Mind as a Function of the Organism*, 102.

57. Lewes, *Problems in Life and Mind: Mind as a Function of the Organism*, 37.
58. Lewes, *Problems in Life and Mind: Mind as a Function of the Organism*, 487.
59. George Lewes, 'The Course of Modern Thought', *Fortnightly Review*, New Series 21 (1877), 321.
60. George Lewes, *Problems in Life and Mind: The Foundation of a Creed* (London: Trübner, 1874), 110.
61. James Sully, 'The Aesthetics of Human Character', *Fortnightly Review*, New Series, 9 (1871), 505.
62. James Sully, *Sensation and Intuition: Studies in Psychology and Aesthetics* (London: Henry S. King, 1874), 15.
63. Sully, *Sensation and Intuition*, 286.
64. Sully, 'The Aesthetics of Human Character', 516.
65. William James, 'On Some Omissions of Introspective Psychology', in Gerald E. Myers (ed.), *Writings, 1878–1899* (New York: Library of America, 1992), 991.
66. James, 'Spencer's Definition of Mind', 894.
67. James, 'On Some Omissions of Introspective Psychology', 991–92.
68. James, 'On Some Omissions of Introspective Psychology', 1002.
69. James, 'On Some Omissions of Introspective Psychology', 1010.
70. Jill M. Kress, *The Figure of Consciousness: William James, Henry James and Edith Wharton* (New York: Routledge, 2002), 28.
71. Kress, *The Figure of Consciousness*, 41.
72. Henry James, *The Portrait of a Lady* (Oxford: Oxford University Press, 2008), 31.
73. James, *The Portrait of a Lady*, 67.
74. James, *The Portrait of a Lady*, 160.
75. James, *The Portrait of a Lady*, 171.
76. James, *The Portrait of a Lady*, 69.
77. Philip Sicker, *Love and the Quest for Identity in the Fiction of Henry James* (Princeton: Princeton University Press, 1980), 56.
78. James, *The Portrait of a Lady*, 222.
79. James, *The Portrait of a Lady*, 223.
80. Sicker, *Love and the Quest for Identity in the Fiction of Henry James*, 56.
81. James, *The Portrait of a Lady*, 52.
82. James, *The Portrait of a Lady*, 48.
83. James, *The Portrait of a Lady*, 185.
84. James, *The Portrait of a Lady*, 114.
85. James, *The Portrait of a Lady*, 103.
86. James, *The Portrait of a Lady*, 135.
87. James, *The Portrait of a Lady*, 137.
88. James, *The Portrait of a Lady*, 251.
89. James, *The Portrait of a Lady*, 378.
90. James, 'Spencer's Definition of Mind', 908.
91. Kress, *The Figure of Consciousness*, 29.
92. James Sully, *Outlines of Psychology: with Special Reference to the Theory of Education*, 2nd edn (London: Longman's Green, 1885), 3.
93. Sully, *Outlines of Psychology*, 48, 55.
94. Sully, *Outlines of Psychology*, 19.
95. Sully, *Outlines of Psychology*, 4–5.

96. Sully, *Outlines of Psychology*, 209.
97. Lewes, *Problems in Life and Mind: The Foundation of a Creed*, 133.
98. Lewes, *Problems in Life and Mind: Mind as a Function of the Organism*, 17.
99. Lewes, *Problems in Life and Mind: Mind as a Function of the Organism*, 152.
100. Lewes, *Problems in Life and Mind: Mind as a Function of the Organism*, 136.
101. Lewes, *Problems in Life and Mind: Mind as a Function of the Organism*, 487.
102. Cline, *The Letters of George Meredith*, 2:731.
103. George Meredith, *Diana of the Crossways: A Novel* (Detroit: Wayne State University Press, 2001), 61.
104. Meredith, *Diana of the Crossways*, 158.
105. Meredith, *Diana of the Crossways*, 225.
106. Meredith, *Diana of the Crossways*, 158.
107. Meredith, *Diana of the Crossways*, 125.
108. Sully, *Outlines of Psychology*, 19.
109. Meredith, *Diana of the Crossways*, 123.
110. Meredith, *Diana of the Crossways*, 124.
111. Meredith, *Diana of the Crossways*, 124.
112. Meredith, *Diana of the Crossways*, 125.
113. Meredith, *Diana of the Crossways*, 126.
114. Meredith, *Diana of the Crossways*, 136.
115. Meredith, *Diana of the Crossways*, 323–24.
116. Lewes, *Problems in Life and Mind: Mind as a Function of the Organism*, 137.
117. Meredith, *Diana of the Crossways*, 189.
118. Meredith, *Diana of the Crossways*, 225.
119. Gillian Beer, *Meredith: A Change of Masks. A Study of the Novels* (London: Athlone, 1970), 97.
120. See Leary, *Metaphors in the History of Psychology*, 332.
121. Rylance, *Victorian Psychology and British Culture*, 12.

6
Alfred Wallace's Conversion: Plebeian Radicalism and the Spiritual Evolution of the Mind

Iain McCalman

Ten years after the publication of *On the Origin of Species*, when Charles Darwin and his disciples were chalking up a succession of victories in the internecine wars over evolution, they were unexpectedly raked by friendly fire. The last few paragraphs of a geology review by Alfred Russel Wallace, co-discoverer of the theory of natural selection and one of its most trenchant interpreters, stated that natural selection could not actually explain the evolution of man's mind. He thought it more plausible that a higher 'Intelligence' had guided the growth of consciousness to specific ends.[1] Darwin, forewarned by Wallace to expect something unusual, had expressed the joking hope that the review would not murder 'too completely your own and my child'.[2] In fact, the real infanticide was still to come.

In a lengthier paper of 1870, entitled, 'On the Limits of Natural Selection as Applied to Man', Wallace followed up his incendiary review by recasting an idea he'd first advanced six years earlier in a much-praised paper 'On the Origin of Human Races and the Antiquity of Man'. Here, he'd argued that 'a grand revolution in nature' had taken place in human evolution, probably during the Miocene period, when the mind had evolved consciousness. Up to that point, he claimed, man's body and brain had been co-shaped by the inexorable struggle for survival that nature visited on all species. The advent of consciousness had deflected the forces of evolution away from the human body, which could now combat environmental changes with tools, clothes and crops, and redirected its transformative energies onto the sphere of minds and morals. He predicted that progressive mental improvements would continue until man once again became a unified species and life 'a bright paradise'.[3]

Wallace's new paper of 1870, 'On the Limits of Natural Selection', imposed a second historical break on human evolution, and one that severed the growth of the conscious mind altogether from the implacable forces of natural selection. Adopting the stock Darwinian dictum that naturally selected change can occur only if beneficial to the variety or species involved, Wallace contended that this process could not therefore have produced the large brain of primitive man. Such a massive brain was unnecessary at a time when primitive man depended primarily on animal attributes of stealth, strength, speed and cunning. It could only have begun to fulfil its potential much later when sociable organization generated a desire for abstraction, aesthetics and ethics. In the absence of a better hypothesis Wallace could only conclude that 'a superior intelligence has guided man in a definite direction, and for a special purpose, just as man guides the development of many animal and vegetable forms'.[4] Wallace also doubted Darwin's assumptions about the material origins of conscious life. 'Greater and greater complexity, even if carried to an infinite extent, cannot of itself, have the slightest tendency to originate consciousness in . . . molecules or groups of molecules. . . . You cannot have in the whole what does not exist in any of its parts'.[5]

Charles Darwin was devastated by this retreat. He scored Wallace's articles with triple underlining and marginal notations of 'no, no!'.[6] In a pained letter to Wallace he denied that it was necessary to introduce higher intelligences to explain man's mental descent from animals: small incremental gradations of evolutionary complexity from primate to man over eons of time were enough to account for consciousness, premonitions of which had been noted in both his and Wallace's previous investigations of animal behaviour. Darwin could hardly believe Wallace's 'metamorphosis'; thinking about it, his pen and lips groaned in unison, 'eheu, eheu'.[7] In reply, Wallace begged his mentor to suspend judgement as to his insanity and to make a fair investigation of his claims. He himself had in the last few years examined with all the scientific rigour he could muster a variety of evidences advanced in support of the new American-imported movement of Spiritualism. Though sceptical at first, he was now persuaded that immaterial beings existed, had the capacity to influence matter, and could advance human mental and moral progress.[8]

Darwin didn't call Wallace mad – at least not in public. He was too grateful to the younger man for his generous response to their coincident announcement of natural selection in 1858, and too admiring of his contributions to natural history to certify him. Other scientific colleagues were less charitable. With hindsight, we can see that Wallace's so-called 'spiritualist conversion' marks the point in his career when he became unrespectable. True, some 'creationists', as Thomas Huxley dubbed them, welcomed Wallace's apparent backsliding on mental evolution, but they remained discomfited and annoyed by his outspoken criticisms of Christianity and his vehement support for natural selection in all other spheres of nature.

On the other side, Wallace's Darwinist colleagues began to regard Wallace as a heretic. Thomas Huxley, once a keen supporter, wrote a blistering rejoinder to the 1870 paper, attacking fallacies in its understanding of materialism as well as inconsistencies with Wallace's earlier claims for the sophistication of 'savage' cultures.[9] Darwin's closest friend Joseph Hooker, a rising young botanist, shifted from admiration of Wallace to open disdain: he told Darwin that the man had 'lost caste' by his embrace of crankish social ideas.[10] Wallace's litany of faddish causes was seen to be mounting: within the next decade it extended to socialism, land nationalization, phrenology, phreno-mesmerism and Anti-Vaccination. George John Romanes, an Oxford-trained expert on mental evolution, inaugurated an enduring tradition of separating Wallace the explorer-naturalist from Wallace the social philosopher. One was 'a man of science', the other 'a man of nonsense'.[11] Wallace noted ruefully in the 1880s that he was still sometimes treated with respect as a biologist, but invariably as a 'babyish idiot' in all other spheres of his thought.[12]

Modern historians have often followed suit, either glossing over Wallace's 'heretical' social ideas or excusing their eccentricity as a sad by-product of a man too credulous for his own good. Wallace's continuous history of eclipse is enshrined in the successive titles of modern biographers: *Darwin's Moon; In Darwin's Shadow; The Heretic in Darwin's Court*.[13] In recent decades, however, a series of revisionist articles, mainly in the *British Journal for the History of Science*, have gone some way towards challenging posterity's 'enormous condescension'. By careful attention to Wallace's social as well as scientific ideas, including his involvements with Owenite socialism, phrenology and spiritualism, the revisionists question conventional separations between Wallace's scientific and social thought and defend the internal logic of Wallace's 'eccentric' positions. John R. Durant, for example, argues that Wallace converted to spiritualism because of 'enduring contradiction between his philosophy of nature [i.e., natural selection] and his philosophy of man and society [spiritualism and socialism]'.[14] This is to say, the hereditarian, materialist and mechanistic principles of natural selection 'came into conflict' with Wallace's reformist social aspirations. Durant and others also make the valid point that sympathy for spiritualism was not necessarily as eccentric in the 1870s as it might now seem.[15]

These are valuable revisions, yet they leave some problems. Why did young Wallace become an Owenite radical and a phrenologist in the first place? Why did he gravitate towards these specific heterodox social movements, and what did he hope to get out of them? How, moreover, do we reconcile his belief in the secular ideologies of Owenism and phrenology with his subsequent adoption of spiritualism? Finally, how do we explain the precise timing of Wallace's 1869–70 conversion, made only a few years after his endorsement of Darwin's naturalistic explanation of the evolution of human consciousness? In other words, what was the 'common social context' that made his 'spiritualist conversion' possible?

Just how fruitful it can be to search for the 'common context' of intellectual shifts like these is evident in an article published a decade ago by the distinguished Darwin biographer James Moore. Moore explores the genesis of Wallace's famous 'Malthusian moment' of February 1858 in the Malay Archipelago. Why, Moore asks, did Wallace 'happen', while on the remote island of Gilolo, to remember Thomas Malthus's essay on population checks, and thus arrive – like Darwin before him – at the mechanism to explain the evolutionary origin of species? Moore's answer is that Wallace's socio-geographic context of 1858 brought to mind an analogous moment twenty years earlier when he was working as a surveyor among the small tenant farmers of South Wales during the popular protest campaigns known as the Rebecca Riots. Faced with comparable circumstances on the islands of Gilolo and Ternate, Wallace remembered reading Malthus's grim thesis on population checks that had seemed so pertinent to 1840s Wales. Having transposed Malthus's theory from South Wales to South East Asia and from man to animals, he scribbled the essay 'On the tendency of varieties to depart indefinitely from the original type', which he then tossed like a grenade into Darwin's secluded life.[16]

In this paper I want similarly to explore the 'common context' of Alfred Wallace's adoption of a spiritualist interpretation of man's mental evolution. Existing explanations of this 'conversion' are inclined to fall down in two related ways: first, they treat the question as a study in the abstract history of ideas rather than as a holistic problem of social and cultural history; secondly, they tend to measure Wallace's intellectual trajectory against that of his generational cohort of Darwinists such as Thomas Huxley and Joseph Hooker. It is true that these young scientists had much in common with him, including having each undertaken formative naturalistic voyages in the southern hemisphere.[17] But by upbringing and – still more – by formal scientific training, they differed sharply from Wallace. In 1870 both Huxley and Hooker were well on their way to becoming leaders of a formidable new cadre of middle-class scientific professionals; Wallace, however, remained by his own account a permanent amateur. To understand his 'common context' of 1870 we must situate his life and career firmly within the culture that shaped him – that of a self-educated plebeian radical of early-Victorian Britain. Once we do this, Wallace's 'spiritualist conversion' becomes both more understandable and more representative.

I

Wallace's money-strapped boyhood and scanty education are well known.[18] The son of a dreamy country solicitor who dragged himself and his family into poverty, Alfred was the second youngest boy out of nine children, more than half of whom died before adulthood. If ever there was a literal embodiment of the Malthusian calculus, it was the Wallace family. As they shifted from place to place in an effort to outpace dwindling resources, Alfred scratched together a rudimentary education at a small Hertford grammar school before stopping formal schooling at fourteen. By this time the shy, gawky boy had endured the humiliations of having to wear homemade sleeve-covers to protect his fading clothes and to work as the school's sole pupil teacher in exchange for fee reductions: '[I]t exposed me,' he later wrote, 'to what I thought was the ridicule or contempt of the whole school'.[19] When his father died in 1844, the family's gradual proletarianization was completed; Mrs Wallace became a domestic servant and the four boys were all apprenticed to mechanic trades.

By taking this step, though, Wallace found himself within a milieu and culture that he could embrace with some pride by comparison with his previous hapless social position. He began his mechanic career by tramping to London with his brother John, who had been offered an apprenticeship within the booming building trades. Not long after this, Alfred was apprenticed briefly to a watchmaker, before joining up with his oldest brother, William, to learn the trade of surveying. All Alfred's subsequent knowledge of science was gleaned outside the orthodox knowledge system. He learnt trigonometry, geometry, nautical astronomy and elements of geology from practical lessons and stray books supplied by his brother. Rambling through the countryside in slack times, he taught himself the rudiments of the Linnean system of botanical orders from a shilling pamphlet produced by the Society for the Diffusion of Useful Knowledge. Later he received his first instruction in entomology from an amateur enthusiast, Henry Bates, whom he befriended by chance at the Leicester public library. The consolatory fascinations of natural history seem to have gripped many such members of the British industrial working classes. For the first time Wallace experienced 'the joy which every discovery of a new form of life gives to the lover of nature'.[20]

His autobiography charts the mental growth of an early-Victorian autodidact. Insatiably curious, he dashed down every intellectual trail that came his way, ranging unsystematically through literature, science, travel, philosophy and mathematics. As a haphazard product of the Utilitarian late-Enlightenment, he soaked up an unshakeable belief in empirical reason and natural laws, a passionate commitment to intellectual self-determination and a pugnacious scepticism of received authority. Like so many British plebeians of the time, he picked up smatterings of technology, science and literature at a variety of Mechanics' Institutes intended to inculcate workmen with self-improving knowledge. While living in the industrializing South Wales town of Neath, he even helped his brother John to design and build a Mechanics' Institute – it still stands today as a symbol of his practical versatility. It was here, too, that he first read Robert Chambers' *Vestiges of the Natural History of Creation*, a colourful bestseller that appealed to his artisan-like delight in demystifying official knowledge. Most importantly it inspired the young surveyor to an ardent belief in 'the concept of evolution through a natural law'.[21] While more systematic naturalists like Darwin viewed the *Vestiges* as a farrago of speculation, Wallace, with all the effrontery of the autodidact, decided that he should now set out to solve the mystery of the origin of species. The book gave him the final grounding for his astonishing decision at the age of 24 to travel to the Amazon basin to work as a self-funded commercial collector of flora and fauna – on the very bottom rung of British naturalism.[22]

Backgrounds like his also constituted a typical recruiting ground for popular radical and freethinking movements. Alfred and his brothers grew to adulthood during the 1830s and 1840s when the Chartist democratic reform movement was at its most militant and when Owenite trades-union and cooperative societies flourished all over the countryside. In 1839, at the age of seventeen, Alfred had attended the Owenite Hall of Science at John Street. He enjoyed the warm sociability, heard Robert Owen himself lecture on cooperation and soaked up the ideology of radical freethought, a movement which saw itself leading a popular democratic struggle against the institutions and philosophies of the established Church. Militant John Street lecturers like the Birmingham mechanic Frederick Hollick preached that organized religion was an elite conspiracy to corrupt and control the people. Radical self-improvement, in both its individual and collective forms, was the philanthropic mill-owner Robert Owen's chief prescription for intellectual and social emancipation. It was a heady and heroic mission. The Hall of Science had been built with the support of William Devonshire Saull, a London wine-merchant, amateur geologist and disciple of Britain's most intransigent republican and 'infidel', ex-tin-maker Richard Carlile. By Wallace's time Carlile had already chalked up nine years of imprisonment in defence of freethinking ideas. Though not himself a disciple of Owen, Carlile shared the John Street podium with a variety of celebrated Owenite lecturers like Benjamin Wardle, John Stevens and Spencer Timothy Hall. All were exponents of 'infidel science', a loose body of sceptical ideas designed to foster an anti-supernatural philosophy of man against the common enemy of 'Priestcraft'. Wallace read Tom Paine, Constantin Volney and Baron Holbach, probably in Carlile's cheap editions, and these works, along with a polemic by Robert Dale Owen against the psychic terrorism of hellfire preachers, persuaded the boy to a lifelong hostility to Christianity.[23]

Historians superficially acquainted with Owenism tend to treat the movement solely as a vehicle for utopian anti-capitalist ideas based on the ethic of cooperation preached by Owen. Owenism was this certainly, but it was also much more. As the movement's greatest historian J.F.C Harrison has argued, it can also be understood as a type of rationalist religion shot through with strains of millenarianism. As well as having cooperative economic agendas, Owenism provided a means for self-improving plebeians to embrace an independent culture with a democratic epistemology. Harrison's profiles of more than ninety Owenite leaders indicate that the movement nurtured multiple sub-ideologies, including those of popular phrenology, phreno-mesmerism and an assortment of fringe healing causes. During the 1830s many Owenite lecturers also took out Unitarian licences to protect themselves from legal prosecution, then opened secular chapels from which they preached varieties of militant freethought. Rank-and-file members seem to have found the format and culture of a popular sect deeply congenial. Like most inheritors of popular Enlightenment traditions, Owenism also managed to accommodate both hard-line materialists and Behemenite mystics steeped in the traditions of radical vitalism that had illuminated 1790s' artisans like William Blake.[24]

Young Alfred Wallace initially encountered phrenology at the John Street Hall of Science and he was inspired as a result to read one of

the best-selling books of the century, George Combe's *The Constitution of Man*, first published in 1828. The new popular science it espoused originated in the theories of a Viennese physician, Franz Gall, and his sometime colleague, J.G. Spurzheim. Combe, however, reshaped their ideas to suit the liberal self-improving mores of early-Victorian Britain. Phrenology advanced the controversial proposition that the brain was the seat of the mind made up of discrete, localized faculties from which sprang all of man's animal propensities, sentiments and intellectual abilities. These faculties reflected, and could be read by, one's topography of cranial bumps, yet in Combe's version these faculties were not absolutely fixed – individuals could inhibit or cultivate them to foster moral behaviour and enhance the intellect. Phrenology could thus be adapted to foster self-improving and even utopian aspirations.[25] Wallace quickly acquired a small phrenological bust and learnt by heart its relief map depicting the seats of emotional and intellectual behaviour.

Around 1843, while teaching for a period at a small Collegiate School in Leicester, Wallace extended his expertise by attending Mechanics' Institute lectures delivered by Spencer Timothy Hall. Hall, an Owenite, was also an advocate of a brand new scientific vogue, phreno-mesmerism.[26] This grafted Combe's calculus of the mind onto the 'animal magnetic' theories of another Viennese physician, Anton Mesmer. He'd postulated the existence of an invisible effluvium circulating through all matter that could be manipulated by human agents to combat mental and physical ills and to enhance well-being. The intellectual amalgam of phrenology and mesmerism tempered Combe's science, not least by extending its range of operations from the mind to the body, which was seen as capable of receiving intellectual and emotional transfers.[27]

Soon Wallace began using his model cranium to diagnose character and practise mesmeric suggestion. By touching a particular spot on his model bust, Wallace believed he could engender corresponding behaviours in his pupils, including producing catalepsy of the limbs and emotional delirium. 'These experiments,' he later wrote, '. . . convinced me, once for all, that the antecedently incredible may nevertheless be true; and, further, that the accusations of imposture by scientific men should have no weight whatever against the detailed observations . . . of other men, presumably as sane and sensible as their opponents'.[28] The sneers and proscriptions of medics and clerics only hardened his distrust of professional authorities. Phreno-mesmerism thus supercharged Wallace's radical Owenite beliefs. It offered him a science of human behaviour that linked the bodies, minds and environments of man in law-like relationships, yet also allowed room for elements of irrationalism, or at least of sentience and emotion. Like Owenism, too, phreno-mesmerism had sect-like aspirations: Roger Cooter, the historian of the movement, goes so far as to call it 'a Methodist science'.[29]

The seven years that Wallace spent during the 1840s surveying the southern English and Welsh countryside in preparation for railways and land enclosures pushed his radicalism in new directions. A Birmingham popular newspaper containing a long amateur poem in praise of Lord Byron introduced Wallace to a new radical hero. Half a century later he could still cite the verses by heart, a paean of praise to a victim of social persecution who had satirized landlords, clergymen and politicians.[30] At the Neath Mechanics' Institute, built by the Wallace brothers in 1846–7, Alfred paid for two separate character readings by itinerant phrenologists. Each diagnosed his fascination with nature, his love of theorizing and his contradictory mixture of diffidence, compassion and pugnacity.[31] Having to survey for enclosures and collect fees from the struggling Welsh-speaking tenant farmers around Neath brought out both his sympathy and his pugnacity.[32] He was moved as a result to attempt his first piece of journalism, an analysis of the South Wales peasantry, whom he thought to be hardy and generous but ill-prepared to survive galloping industrial and agricultural change.[33]

When a year or two later Alfred Wallace came to embark on his naturalist voyages to the Amazon Basin he thus did so as a thoroughgoing Owenite radical. We catch glimpses of this even among the few records of the expedition that survived his shipwreck on the way home to England in 1852.[34] In the Amazon town of Barra, for example, he and his younger brother Herbert performed phreno-mesmeric experiments on the street urchins to prove that 'the people's science' had universal applicability. Owenite beliefs in the universal rationalism of man informed his first encounters with the head-hunting Indian tribespeople of Huapes. Unlike Darwin, who'd been shocked by the savagery of Patagonians, Wallace thought the Huapes had developed an ethical and cooperative lifestyle that harmonized perfectly with their tropical environment.[35] In Borneo on the Malay Archipelago in 1855, he took the typically Owenite view that most Orang-Utan behaviour was socially learned rather than instinctive. Similarly, when analysing the ethnographic differences between Papuans and Malays on the islands of Aru, he stressed Owenite forces of environmental determinism, the plasticity of human character and the shaping restraints of natural boundaries. On the Malay islands of Gilolo and Ternate in 1858, it was his Owenite radical recollection of the Welsh peasantry's struggles that spurred his feverish mind to come up with the idea of the evolution of species by natural selection.[36]

II

On settling back in England in 1862, after eight years in the Malay Archipelago, Wallace found that his stocks had changed. He'd left as a plebeian collector and he returned a scientific celebrity. Colleagues feted his intellectual generosity, fertile theorizing, lucid writing and mammoth collecting achievements. Despite his shyness, he was quickly assimilated into Darwin's inner circle of correspondents and supporters, alongside two other recent Southern hemisphere travellers, Thomas Huxley and Joseph Hooker. 'My good knight', as Darwin soon dubbed him, was a perfect foil to his older mentor – as quick to publish as Darwin was slow, and as fond of controversy as Darwin was of peace. Wallace's rapid succession of scientific papers of the early 1860s carved out new roles for natural selection in the fields of avian and butterfly protective mimicry, the evolutionary origins of man, and the ethnography of Malay Archipelago peoples. This period of the early 1860s also saw Wallace thinking in an unusually conservative political vein: he'd come under the influence of the social theorist Herbert Spencer, whose advocacy of economic laissez-faire and British imperial dominance was based on analogies with the struggle of biological organisms in nature.[37]

Yet Wallace's surface reintegration into British society belied an inner unease. With his thin, stooped body, tiny wire-framed glasses and shaggy white beard, he brings to mind Rip Van Winkle, a man reawakened after a lengthy sleep to a habitat he no longer understood or liked. For twenty-two out of the previous twenty-four years he had lived a tough, solitary life in tropical jungles on the other side of the globe, mostly in the company of native peoples. He'd returned to Britain at the age of thirty-nine to live with his elderly mother, nursing a body afflicted with fever, ulcers and tropical boils. His natural shyness crusted into near agoraphobia. Even sympathetic colleagues like Huxley unnerved him with their professional self-assurance and specialist abilities. Unable to match either their social connections or their technical language, he repeatedly failed to gain a scientific job. As his income dwindled he relied on the hackwork of school examining and journalism. His social gaucherie also frightened off a newly acquired fiancée, the first woman he'd ever loved. Bewildered and hurt, he poured out his heart in a letter to Darwin, who could only offer his usual remedy – immerse yourself in work.[38]

During Wallace's long absence, the political and economic landscape of Victorian Britain had changed as fundamentally as its social mores. He was confronted with industrial and urban conglomerations beyond his wildest imagining. The reform movements of Chartism and Owenism, which had inspired his youthful idealism, had been broken, leaving a vacuum filled by prosperous middle-class industrialists, financiers and professionals. A resultant sense of alienation from conventional professional norms began to show in his writings, especially in a series of papers of the mid-1860s delivered to the London Anthropology Society. Here, debates on race and ethnology – volatile enough at any time – reached new levels of intensity against the backdrop of divided English affiliations over the American Civil War. Wallace angered the mainly conservative members by defending the capabilities of the black races, extolling the 'progress of civilisation in the Northern Celebes' and criticizing the brutality of Australian settlers towards Aborigines.[39] The President of the Society, a brash young physician, James Hunt, took the lead in jeering at Wallace as an unscientific dreamer.[40] By the end of the 1860s Wallace was criticizing the timidity and narrow-mindedness of many professional scientists and expressing his unease at the reactionary social uses to which Darwinian theory was being put. He was thus the earliest of Darwin's disciples to doubt whether the contemporary capitalist world encouraged the survival of those who were ethically fittest. In mid-Victorian Britain it seemed to him that the mediocre or the brutal were more likely to be winners.

These resurgent radical views marked Wallace's abandonment of the laissez-faire sociology of Herbert Spencer and return to the ideas of his original radical guru, Robert Owen. It was now, as he writes in his autobiography, that he came to read Owen's life story in detail and to absorb fully his famous communitarian treatise of 1813, *A New View of Society*. At the same time Wallace could not fail to notice that the decline of Owenism had been accompanied by the emergence of new but kindred social movements, a development that accorded with one of his earliest treatises on evolution. In 1855 he had written a brilliant paper from Sarawak in Borneo postulating that 'every species has come into existence coincident both in space and time with a pre-existing allied species'.[41]

One of these new allied species resembled more a loose and informal conjunction of movements that historians have variously called 'fringe medicine', 'democratic healing' or 'Physical Puritanism'. The last term was coined by a radical Edinburgh doctor named Samuel Brown in the year of Wallace's return from the Amazon. Writing in the *Westminster Review* of 1852, Brown proclaimed that:

> A new sort of puritanism has arisen in our times, and its influence is as extensive as its origin is various. [...] It is a puritanism of the body; and it comes before the world in many names; but the common purpose of all its manifestations is the healing, cleansing and restoration of the animal man.

Its manifestations, he said, included hydropathy, homeopathy, mesmerism, phrenology, herbalism, vegetarianism and teetotalism. Collectively these medical movements constituted what he called 'a rooted and far-spreading conspiracy against orthodoxy . . . risen upon us, suddenly and simultaneously, like an insurrection of citizens against a tyranny grown beyond all endurance'. The chief sources of the tyranny, Brown asserted, were orthodox scientific and medical professionals fixated on drug-centred and blood-letting regimens. Their policing powers and legal exclusiveness had grown out of proportion to their therapeutic abilities.[42]

Brown sensed that this diffuse popular medical movement had caught up a raft of counter-cultural aspirations. Many of the new popular healing regimens, such as Samuel Thompson's botanicals and Sylvester Graham's vegetarianism, had invaded the country from the freer health environment of democratic America. Sect-like in structure and culture, these scientific-cum-metaphysical groupings offered plebeians the power to control their own bodies and minds through an assortment of cheap, naturalistic remedies and practices. Brown believed that Physical Puritanism offered ordinary Britons a complete alternative philosophy of life. He traced its lineage through the Essenes' ancient 'physiological heresies', Paracelsus's alchemical healing, Von Helmont's vitalism, Swedenborg's astral mysticism, Mesmer's universal magnetic fluid, and Gall's phrenological map of the mind.[43]

Victorian Physical Puritanism struck a sympathetic chord with former Owenites not least because of a common recruitment from among artisans, shopkeepers and sub-professionals. Especially in the industrializing north, these were the self-improving classes trying to find their way within the vortices of social change. Their experiences and aspirations made them similarly sceptical of professional elites, established religions and orthodox knowledge systems. Owenite converts to various strands of Physical Puritanism included the John Street lecturers Frederick Hollick, John Stevens and Spencer Hall.

The former John Street disciple Alfred Wallace was equally primed for sympathy. He flirted with vegetarianism and embraced Anti-Vaccination, a cause that appealed strongly to the anti-druggist principles of radical plebeians.[44] Towards the end of the 1860s he also declared himself a believer in the burgeoning new popular healing movement of Spiritualism.

Though originally sceptical of what he'd read about the claims of spiritualism, Wallace shared the radical autodidact's willingness to subject all phenomena to the litmus test of reason, particularly when urged to do so by his favourite sister Frances Sims, who had just returned from America an enthusiast of the movement. Frances may also have hoped that her lovelorn brother would benefit from some spiritual solace. Wallace himself, however, claimed his interest in spiritualism to be impeccably scientific. He proposed to investigate spiritualist claims as he would any other naturalistic phenomenon. After attending several séances in 1866 and meeting a persuasive female medium whom he tested repeatedly for signs of fraud, he declared himself convinced. At the very least, he argued in a pamphlet of 1870, the evidences of spiritualists warranted unbiased scientific scrutiny.[45] Shortly after, he published 'On the Limits of Natural Selection as Applied to Man', the paper on the spiritual evolution of mind that brought Darwin and his colleagues such pain.

III

Though Alfred Wallace's espousal of spiritualism as a motor of human mental evolution is usually represented as a startling 'conversion' on the part of a long-time materialist, it would be more accurate to call it an adaptation of his earlier thought. Some historians of Victorian culture have noted that mid and late-Victorian radicals seemed often to have embraced a polymathic unorthodoxy, rather as if political, religious and medical heresies attracted each other. Wallace appears a case in point. However this multiple heterodoxy was not mere faddism or eccentricity as some historians seem to believe: the explanation seems rather that these diverse democratic movements appealed to a common cluster of plebeian needs – for self-improvement, self-respect and self-determination. Wallace could never understand why his belief in the spiritual basis of human mental evolution proved so damaging to his reputation. He regarded himself as a scientist who'd made a small but vital adjustment to the theory of natural selection so as to enhance its larger credibility. Darwin, by contrast, believed that his friend had tried to slaughter the theory they had co-pioneered. Wallace regarded spiritualism as an exciting, logical and scientifically empirical extension of the naturalistic social theories of Owenism, phreno-mesmerism and Physical Puritanism. All three movements had displayed a marked ambiguity on the issue of materialism. Though secular in tendency, each had also nurtured contrary strains that sometimes shaded into outright mysticism or millenarianism. If this blurring of the lines between determinism and free will, and between metaphysical and materialist views of the world, seems eccentric to modern philosophers, it was nevertheless a source of attraction to Victorian plebeians.

Wallace could not fail to be impressed by the fact that Robert Owen, the man whose political and social philosophies he most admired, had himself converted to spiritualism in the 1850s, shortly before he died. A good number of former Owenites and phreno-mesmerists followed suit. Surprising as it might seem, plebeian spiritualism was no less counter-cultural than many predecessor radical movements – as democratic and anti-Christian as Owenism and as participative and spectacular as Phreno-Mesmerism. Many spiritualists believed, too, that they were actually aiding the secularist cause by naturalizing the supernatural. Already attuned to the operation of imponderable forces like mesmerism, gravity and electricity, Wallace saw nothing illogical in the idea of ethereal beings possessed of intelligence, who might guide the evolution of the human mind. The trances of spiritualist mediums seemed to mirror the states of psychic alteration that he'd induced in subjects on both sides of the globe. The refusal of most professional scientists to give the movement a fair trial only exacerbated his conviction of scientific martyrdom. When James Hunt at the Anthropological Society scoffed at phrenology as an utter failure and sniggered at Wallace's spiritualist informants, his plebeian defiance only hardened.[46] If professional scientists like Hunt wanted to exclude him from their ranks, so be it: he was proud to be a popular amateur scientist and a radical outsider.

Above all spiritualism offered Wallace a way to recover his ingrained utopianism in the face of the depressingly reactionary applications of social Darwinism of many colleagues. The new popular spiritualist movement from America appealed to his instinctively optimistic philosophy of life: it exempted humans from the savage operations of natural selection and encouraged them to improve both their environment and themselves. Mental and moral capacities would continue to evolve even after the decay of the body, leading eventually to that bright socialist paradise of which he'd always dreamed.

Notes

1. John. R. Durant, 'Scientific Naturalism and Social Reform in the Thought of Alfred Russel Wallace', *British Journal for the History of Science*, 12 (1979), 45.
2. Ross A. Slotten, *The Heretic in Darwin's Court: The Life of Alfred Russel Wallace* (New York: Columbia University Press, 2004), 288.
3. Alfred Russel Wallace, 'The Origin of Human Races and the Antiquity of Man Deduced from the Theory of Natural Selection' (a paper read at the ASL meeting of 1 March 1864), *Journal of the Anthropological Society of London*, 2 (1864), clviii–clxx (followed by an account of related discussion on pp. clxx–clxxxvii), at: www.wku.edu/~smithch/wallace/S093.htm.
4. Alfred Russel Wallace, 'On the Limits of Natural Selection as Applied to Man', in *Contributions to the Theory of Natural Selection* (London, New York: Macmillan, 1870), 359.
5. Wallace, *Contributions to the Theory of Natural Selection*, 365.
6. Slotten, *The Heretic in Darwin's Court*, 270.
7. Charles Darwin to Alfred Russel Wallace, 26 January 1870, in James Marchant (ed.), *Alfred Russel Wallace: Letters and Reminiscences* (London, New York, Toronto and Melbourne: Cassell and Company, 1916), 1:251.
8. Durant, 'Scientific Naturalism and Social Reform', 45–46.
9. Slotten, *The Heretic in Darwin's Court*, 286–87.
10. Slotten, *The Heretic in Darwin's Court*, 357–58.
11. Durant, 'Scientific Naturalism and Social Reform', 32.
12. Wallace, *Alfred Russel Wallace: Letters and Reminiscences*, 2:225.
13. Michael Shermer, *In Darwin's Shadow: The Life and Science of Alfred Russel Wallace* (Oxford: Oxford University Press, 2004); Slotten, *The Heretic in Darwin's Court: The Life of Alfred Russel Wallace*; Amabel Williams-Ellis, *Darwin's Moon: A Biography of Alfred Russel Wallace* (London: Blackie, 1966).
14. Durant, 'Scientific Naturalism and Social Reform', 51.
15. See Roger Smith, 'Alfred Russel Wallace: Philosophy of Nature and Man', *British Journal for the History of Science*, 6 (1972), 177–99; Malcolm Jay Kottler, 'Alfred Russel Wallace, the Origin of Man and Spiritualism', *Isis*, 65 (1974), 144–92; Charles H. Smith, 'Spritualism, and Beyond: "Change," or "No Change"?' in Charles H. Smith and George Beccaloni (eds), *Natural Selection & Beyond: The Intellectual Legacy of Alfred Russel Wallace* (Oxford and New York: Oxford University Press, 2008), 391–423; Roger Cooter, *The Cultural Meaning of Popular Science: Phrenology and the Organization of Consent in Nineteenth-Century Britain* (Cambridge, London and New York: Cambridge University Press, 1984); Alison Winter, *Mesmerized: Powers of Mind in Victorian Britain* (Chicago and London: University of Chicago Press, 1998), 117–21.
16. James R. Moore, 'Wallace's Malthusian Moment: The Common Context Revisited', in Bernard Lightman (ed.), *Victorian Science in Context* (Chicago and London: University of Chicago Press, 1997), 290–311.
17. Iain McCalman, *Darwin's Armada: How Four Voyagers to Australia Won the Battle for Evolution and Changed the World* (Melbourne: Penguin, 2009).
18. For biographical details on Wallace see McCalman, *Darwin's Armada*; Shermer, *In Darwin's Shadow*; Slotten, *The Heretic in Darwin's Court*; Williams-Ellis, *Darwin's Moon*; and Alfred Russel Wallace, *My Life. A Record of Events and*

Opinions (London: Chapman & Hall, 1908; facsimile copy, Elibron Classics, 2005).

19. Wallace, *My Life*, 33.
20. Wallace, *My Life*, 104.
21. Wallace, *My Life*, 144.
22. Wallace, *My Life*, 144.
23. See Greta Jones, 'Alfred Russel Wallace, Robert Owen and the Theory of Natural Selection', *BJHS*, 35 (2002), 73–96; Wallace, *My Life*, 43–46.
24. J.F.C. Harrison, *Quest for the New Moral World. Robert Owen and the Owenites in England and America* (New York: Scribner, 1969).
25. Cooter, *The Cultural Meaning of Popular Science*, 101–27; for Wallace's later evaluation of Combe, see Alfred Russel Wallace, *The Wonderful Century* (London and New York: Swan, Sonnenschein, 1899), 162–70.
26. On Spencer Hall and the impact of phreno-mesmerism more generally, see Winter, *Mesmerized*, 112–35.
27. Cooter, *The Cultural Meaning of Popular Science*, 150, 175–81.
28. Wallace, *My Life*, 127.
29. Cooter, *The Cultural Meaning of Popular Science*, 194.
30. Wallace, *My Life*, 63.
31. Wallace, *My Life*, 140–41.
32. Wallace, *My Life*, 134.
33. Reproduced in full in the two-volume edition of Alfred Russel Wallace, *My Life: A Record of Events and Opinions* (London: Bell, 1905), 1:206–22.
34. Wallace, *My Life* (1908), 156.
35. Wallace, *My Life* (1908), 150–51.
36. Jones, 'Alfred Russel Wallace, Robert Owen', 76, 80–86; and Wallace, *My Life*, 190.
37. Wallace, *My Life*, 57.
38. Wallace to Darwin, 20 January 1856, and Darwin to Wallace, 29 January 1865, *Darwin Correspondence Project*, at: www.darwinproject.ac.uk/darwinletters/calendar/entry-4750.html - mark-4750.f1.
39. Slotten, *The Heretic in Darwin's Court*, 219–22.
40. Slotten, *The Heretic in Darwin's Court*, 213–15.
41. Shermer, *In Darwin's Shadow*, 85.
42. Samuel Brown, 'Physical Puritanism', *Westminster Review*, 112 (1852), 405–42.
43. Brown, 'Physical Puritanism', 405–42.
44. Wallace, *My Life*, 309, 329–33.
45. Wallace, *My Life*, 335–37.
46. Slotten, *The Heretic in Darwin's Court*, 246–47.

7
Molecular Machines and Lascivious Bodies: James Clerk Maxwell's Verse-born Attacks on Tyndallic Reductionism

Daniel Brown

I

The 1874 Belfast address was the most notorious, but not the first, occasion that the physicist John Tyndall used to present his mechanistic account of the natural world. Tyndall, the Professor of Natural Philosophy at the Royal Institution, introduces his doctrine of 'scientific materialism' to his colleagues in the presidential address he gave to Section A, the Mathematics and Physics section, of the British Association for the Advancement of Science (BAAS) meeting at Norwich in 1868.[1] He argues that atoms are held together by innate forces, so that they cohere as molecules and then as more complex self-organizing aggregates, including plants and animals capable of reproducing themselves and of evolving into various further forms through the Darwinian principle of natural selection. Early in the address he notes a formal correspondence between the Egyptian pyramids and the crystalline aggregates that salt molecules assume, and draws from it the observation that while 'the final form of the pyramid expressed the thought of its human builder' there is no need to invoke a creator in the case of the 'salt-pyramids', for 'these molecular blocks of salt are self-posited, being fixed in their places by the forces with which they act upon each other'. Describing '[t]his tendency on the part of matter to organize itself' as 'all-pervading' and, more provocatively still, as '[i]ncipient life',[2] Tyndall moves by analogy from the crystals of salt to 'a living grain of corn'[3] and from there storms all of organic nature, barely stopping short of explaining human thought, to sketch a scrupulously deterministic cosmology that has no need of a divine Creator and Designer.

The Norwich Address begins provocatively by considering the watch as an analogy for objects in nature, a pointed reference to the foundational argument for Design with which William Paley begins his much reprinted *Natural Theology or, Evidences of the Existence and Attributes of the Deity, Collected from the Appearances of Nature* (1802).[4] In a calculated affront to the Design hypothesis, Tyndall acknowledges that the 'motion of the [watch] hands may be called a phenomenon of art, but,' he writes, 'the case is similar with the phenomena of nature' not on the grounds that they too are instances of artifice that imply a maker, but rather that a force impels them both, for like the watch, natural phenomena 'also have their inner mechanism, and their store of force to set that mechanism going'.[5] Unlike the man-made watch, he argues, objects in nature imply no original creator, for their forms follow from this principle of a radical force. Tyndall's atomic forces are made to do for all matter what Darwin's principle of natural selection achieves only for organic nature: furnish a naturalistic mechanism that effectively inverts the natural theological argument, so that evidence of pattern and design is seen accordingly to demonstrate not the necessity of a Creator but, on the contrary, nature's radical independence from such a principle.

Tyndall and his friend the biologist T.H. Huxley were the de facto leaders of the London-based 'Metropolitan' scientists, who in the 1860s and 1870s gathered around this creed of scientific materialism. Having formally banded together in 1864 as the X Club, the group also included Edward Frankland, John Lubbock, Thomas Hirst, Herbert Spencer, George Busk, Joseph Hooker and William Spottiswoode. The Metropolitans were pitted against another tribe, the 'North British' physicists, a group of Presbyterian Scots that included James Clerk Maxwell, Peter Guthrie Tait, Fleeming Jenkin and William Thomson. Maxwell attacks Tyndall's reductionism obliquely in addresses to the British Association and more playfully and often crudely in a series of comic verses he wrote in the early 1870s for his scientific peers in the Red Lions club, which gathered during the annual meetings of the BAAS for informal and indeed raucous dinners, a carnivalesque counterpoint to the day's serious business of science papers and committee work.

Energy physics, field theory, the theory of gases and other develop-ments in the 1850s and 1860s brought the classical doctrine of atom-ism back into scientific respectability, which led to the North British and Metropolitan tribes battling for proprietary rights to specify the nature of the atom and its consequences. Replying to Tyndall's Norwich address in the corresponding address he gave as sectional president at the Liverpool meeting of the BAAS in 1870, Maxwell challenges scientific materialism by surveying and promoting alternative lines of research that explain the properties of molecules statistically 'on dynamic principles', that is, through a mechanistic ontology of matter and motion. He had demonstrated this approach in his work on the diffusion of gases, in which he theorized the physical properties of gas by analogy with the mechanics of particles, seeing it as consisting of rapidly moving elastic solid particles that constantly collide with one another. Acknowledging the hypothetical nature of atoms and molecules, Maxwell and his peers accordingly discerned the existence and nature of these sub-microscopic particles statistically, through their collective behaviour. Maxwell contrasts the statistical account of molecular behaviours with another, rather tautologous, speculative approach that strongly suggests Tyndall's reductionist hypothesis of a force that binds all matter: 'If a theory of this kind should be found, after conquering the enormous mathematical difficulties of the subject, to represent in any degree the actual properties of molecules, *it will stand in a very different scientific position from those theories of molecular action which are formed by investing the molecule with an arbitrary system of central force invented expressly to account for the observed phenomena*'. Such forces are, he observes, not scientific principles but 'occult properties'.[6]

In contrast to Tyndall's scientific materialism, which as was noted earlier naturalizes mechanistic determinism on a cosmological scale, seeing it to apply to phenomena from atoms and molecules up to the mind that theorizes such schemes, the statistical theory championed by Maxwell and his peers recognizes that molecules have different properties from the bodies they compose and that their mechanistic behaviours can interact in unforeseen ways. This understanding is highlighted by the most famous of Maxwell's 'thought experiments', which he elaborates in his 1871 book *Theory of Heat*, where a hypothetical agent that William Thomson named 'Maxwell's demon', but which its inventor preferred to call simply a 'valve', mediates the flow of gas through an aperture between a container of warmer gas and another of colder gas, to allow warmer molecules in the colder compartment to move to the warmer group, in apparent contravention of the second law of thermodynamics, which asserts that all energy in the form of heat moves to distribute itself evenly throughout space. With his 'demon' Maxwell argues that the second law is not the strict invariant principle that its title suggests but a statistical principle.

Through his extrapolations from the statistical theory, most boldly illustrated by his hypothetical 'demon', Maxwell finds the instabilities and variability commonly identified with the workings of complex organisms and human moral character entrenched at the sub-microscopic level of atoms and molecules. As Thomson recognizes with his appellation of 'Maxwell's demon', by extending this principle of instability to the behaviours of atoms and molecules, the mechanistic relations of physics come to assume an anthropomorphic character. This implication of Maxwell's physical theory, of a radical analogy between human and inorganic behaviours – the opposite of that promoted by Tyndall – furnishes the playful premise for his poems from the early 1870s.

The earliest of this group, 'In Memory of Edward Wilson, *Who repented of what was in his mind to write after section*', is a parody of Robert Burns' song 'Comin' thro the Rye', which questions the need for a body to 'cry', 'tell' or 'gloom' after it has kissed another body it happened to meet. In Maxwell's parody, it is not human bodies that come into accidental contact with one another, but the simple inorganic bodies studied by mechanics: 'Gin a body meet a body / Flyin' through the air'. 'In Memory of Edward Wilson, *Who repented of what was in his mind to write after section*' displaces the sly sexual innuendo of its model, which represents the type of thoughts that Wilson allegedly had during the proceedings of the Mathematics and Physics section of the BAAS, with verses that focus upon the activities of those *inorganic* bodies that are more properly thought about in such meetings. The poem appears to be a version of a parody Maxwell wrote as an undergraduate at Cambridge during the early 1850s, which he revised, or at least re-titled, for a new audience of Red Lions in the early 1870s.[7] The reference to '*section*' in the title is obvious, while that to Edward Wilson names an Irish astronomer and physicist (1851–1908), who accompanied William Huggins to Algeria in 1870 to witness an eclipse. Wilson may have rashly admitted to distracting thoughts during a Mathematics and Physics meeting at the BAAS, or else he was simply a young colleague who could be teased in this way, and sent off like a schoolboy to do such mechanics problems as punishment '*after section*'.

Maxwell's parody springs from its quibbling use of the word 'body', which mobilizes and juxtaposes the starkly contrastive connotations that each of its two referents, the human body and the inorganic body studied by mechanics, acquire when they make physical contact with their own type. Not quite coquettish, the '*Rigid Body*' that '*sings*' here is nonetheless depicted as the flighty object of young men's attentions:

> Gin a body meet a body
>   Flyin' through the air
> Gin a body hit a body,
>   Will it fly? And where?

Ilka impact has its measure,
    Ne'er a ane hae I,
Yet a' the lads they measure me,
    Or, at least, they try.

Gin a body meet a body
    Altogether free,
How they travel afterwards
    We do not always see.
Ilka problem has its method
    By analytics high;
For me, I ken na ane o' them,
    But what the waur am I?[8]

'Flyin' through the air' freely and having no understanding of the problems it gives rise to and how they could be solved, the Rigid Body is depicted as carefree and careless. The world of elementary mechanics is characterized here by phenomenal promiscuity and abandon. The counterpart to the yielding human body of 'Comin Thro' the Rye', the correspondingly rigid and inorganic body of Maxwell's poem does what it wants. An amalgam of mechanism and agency that defies Tyndallic determinism, it furnishes a suggestive model for the clashing atoms of Maxwell's dynamical theory of gas. Like such atoms, the rigid body cannot be measured and indeed is able to move irregularly, 'Altogether free', a model for the sub-microscopic bodies that Maxwell describes with his hypothesis of the 'demon'.

The statistical theory delivers individual atoms and molecules from Tyndall's deterministic account, where, as Maxwell describes it in his Liverpool address, 'the molecules obey the laws of their existence, clash together in fierce collision, or grapple in yet more fierce embrace, building up in secret the forms of visible things'.[9] He offers a contracted but unexpurgated version of this account of Tyndall's minute world a year later in 'A Tyndallic Ode'. The poem's persona discloses his identity through a mock-parlour game riddle:

I come from empyrean fires—
From microscopic spaces,
Where molecules with fierce desires,
Shiver in hot embraces[10]

Sexual lust furnishes a naturalistic metaphor for Tyndall's deterministic materialism. The self-organizing 'tendency' of matter that he makes indigenous to his region is recognized in Maxwell's poem as a mythic extrapolation from sexual acts. The dubious sexual associations and interests which, as Gowan Dawson documents, coloured Tyndall's reputation from the 1860s are invoked here to casually characterize his entire cosmology, a presumptive identification of a metaphysical materialism with licentious hedonism that Walter Pater would also be subjected to in subsequent decades.[11]

Tyndall's scheme is characterized alike in Maxwell's poem and his Liverpool address by the conflation of atomist physics and biological evolution, for the 'laws of their existence', the atoms clashing together and their complementary 'fierce embrace' and 'hot embraces', suggest the twin Darwinian exigencies of the competitive struggle for survival and the sexual drive to reproduce. Hence the satirical label that Maxwell gives for Tyndall's encompassing doctrine, 'Molecular Evolution'.

The first of his poems that share this title was published in *Nature* in October 1873, having been written in September of that year during the Bradford meeting of the BAAS, while the second, to which he helpfully gives the alternative title 'Song of the Cub', dates from the subsequent meeting at Belfast in August 1874, it being one of several poems he wrote in response to Tyndall's Presidential Address.[12] The word *evolution* gained currency in the 1860s through not only the public controversy that followed the publication of Darwin's *Origin of Species* in 1859 but also a revival of interest in Laplace's nebular hypothesis, which maintains that the universe itself evolved as the condensation of gases. The two applications of the term are synthesized and popularized by Herbert Spencer's philosophical system, which elaborates evolution into a universal principle.[13] In his 1873 poem Maxwell parodies what he sees as Tyndall's illegitimate extrapolation of the physical universe from his hypothesis of atomic forces with a parodic cosmology that he develops from his own premises. To do this he invokes Lucretius's great Latin poem on atomism, *On the Nature of Things*.

II

'Molecular Evolution' opens with the Lucretian account of the creation of the universe:

> At quite uncertain times and places,
>   The atoms left their heavenly path,
> And by fortuitous embraces,
>   Engendered all that being hath.[14]

The Lucretian story begins with the atoms in 'their heavenly path', raining down in a laminar flow that is finally disrupted by the clinamen, a slight swerving of some atoms, the effect of which is compounded as they collide with others and rebound to cause further collisions, and through them aggregations that form things, indeed whole worlds.

H.A.J. Munro's critical edition of Lucretius, which appeared first in 1860 and from 1864 with a supplementary volume of translation and further commentary, was, as Frank M. Turner observes, decisive in shifting the prevailing Victorian characterization of Lucretius from poet to scientist.[15] The influential review essay of Munro's second edition (1866) that Fleeming Jenkin wrote for the *North British Review* in 1868, 'The Atomic Theory of Lucretius', similarly works to restore the poet's scientific reputation by comparing his ideas with recent advances in physics; 'we may profitably consider what the real tenets of Lucretius were, especially now that men of science are beginning, after a long pause in the inquiry, once more eagerly to attempt some explanation of the ultimate constitution of matter'.[16] While Maxwell knew Lucretius's text from his early studies in Latin and philosophy,[17] the phrase he uses for the clinamen in the first line of his poem echoes those that Fleeming Jenkin employs in his article on Lucretius, 'at quite uncertain times and uncertain places', and cites from Munro's translation, 'at quite uncertain times and uncertain points of space'.[18] Maxwell corresponded with Jenkin about his essay on Lucretius,[19] and it is likely that the two discussed the Roman poet through their reading of Munro. Maxwell's phrase 'at quite uncertain times and places', which appears in many of his writings from the late 1860s and early 1870s, indicates the central importance of Munro's edition for his renewed consideration of Lucretius at this time. Indeed it is probable that Munro used such a formulation in replying to a letter Maxwell wrote to him in early 1866, asking about the precise nature and behaviours of atoms in Lucretius.

Maxwell was drawn to Lucretius's scientific speculations through his work on the kinetic theory of gases. He begins his 1866 paper 'On the Dynamical Theory of Gases' by tracing precursors to this theory. Less sure of the anticipations of the theory he discerns in Lucretius than those he has traced in the history of modern physics, Maxwell writes to Munro for advice: 'With respect to those who flourished since the revival of science I can make out pretty well what they really meant but I am afraid to say anything of Lucretius because his words sometimes seem so appropriate that it is with great regret that one is compelled to cut off a great many marks from him for showing that he did not mean what he has already said so well.'[20] Munro's reply to this letter appears not to have survived, but, as the published version of 'On the Dynamical Theory of Gases' demonstrates, it evidently vindicated Maxwell's hopes, allowing Lucretius's speculative doctrine to lend its ancient authority to the North British ontology of matter in motion:

> The opinion that the observed properties of visible bodies apparently at rest are due to the action of invisible molecules in rapid motion is to be found in Lucretius. In the exposition which he gives of the theories of Democritus as modified by Epicurus, he describes the invisible atoms as all moving downwards with equal velocities, which, at quite uncertain times and places, suffer an imperceptible change, just enough to allow of occasional collisions taking place between the atoms. These atoms he supposes to set small bodies in motion by an action of which we may form some conception by looking at the motes in a sunbeam. The language of Lucretius must of course be interpreted according to the physical ideas of his age, but we need not wonder that it suggested to Le Sage the fundamental conception of his theory of gases, as well as his doctrine of ultramundane corpuscles.[21]

This passage includes Maxwell's earliest use of the phrase 'at quite uncertain times and places', which, as was suggested earlier, may indicate Munro's use of it in the reply he wrote to Maxwell's letter. Occupying the opening line of 'Molecular Evolution' and occurring insistently at many points in Maxwell's later writings as well as in Fleeming Jenkin's essay, the phrase is central to the North British reading of Lucretius, for on it turns their polemical use of his poem as a defence of free will against the mechanistic construal of atomism that Tyndall and his peers were promoting. Lucretius's doctrine of human free will is integral to his physics. He presents it as a direct consequence of the clinamen:

> Do you see then in this case that, though an outward force often pushes men on and compels them frequently to advance against their will and to be hurried headlong on, there yet is something in our breast sufficient to struggle against and resist it? . . . admit that besides blows and weights there is another cause of motions, from which this power of free action has been begotten in us, since we see that nothing can come from nothing. For weight forbids that all things be done by blows through as it were an outward force; but that the mind itself does not feel an internal necessity in all its actions and is not as it were overmastered and compelled to bear and

put up with this, is caused by a minute swerving of first-beginnings at no fixed part of space and no fixed time.[22]

Maxwell explains the Lucretian derivation of free will from the clinamen in his 'Discourse on Molecules', which like 'Molecular Evolution' was delivered at the Bradford meeting of the BAAS in 1873. The account that Tennyson gives of the clinamen in his poem 'Lucretius' provides the text for his discussion:

In his dream of nature, as Tennyson tells us, he

> "Saw the flaring atom-streams
> And torrents of her myriad universe,
> Running along the illimitable inane,
> Fly on to clash together again, and make
> Another and another frame of things
> For ever."

And it is no wonder that he should have attempted to burst the bonds of Fate by making his atoms deviate from their courses at quite uncertain times and places, thus attributing to them a kind of irrational free will, which on his materialistic theory is the only explanation of that power of voluntary action of which we ourselves are conscious.[23]

The clinamen may accordingly be said to entrench a principle of free will at the most radical level of physical reality. This implication of the Lucretian account receives an additional emphasis in the draft of Maxwell's *Encyclopaedia Britannica* article on the 'Atom', which Harman dates to September 1874, a year after he presented the 'Discourse on Molecules' and 'Molecular Evolution' to his peers at the Bradford meeting:

According to his description of the doctrine of Democritus the atoms are all in motion in a downward stream with an enormous speed. [. . .] At quite uncertain times and places these atoms are deflected from their vertical path. This causes them to jostle one another and by their fortuitous concourse they form visible bodies.

The final clause in this passage restates the third and fourth lines of 'Molecular Evolution', where the atoms 'by fortuitous embraces, / Engendered all that being hath'. The phrase 'fortuitous concourse' from the prose extract is the English translation of the Latin *concursus fortuitus* that Cicero uses to describe the Lucretian clinamen.[24] In its echo in the third line of the poem this phrase appears to be modified not only to satisfy the needs of the rhyme scheme but also to suggest a parody of sexual activity, so that it offers a counterpoint to the similarly anthropomorphic description of Tyndall's version of atomism in the second stanza of the '*Tyndallic Ode*'. The 'fortuitous embraces' are chance encounters that can be described by the anthropomorphic analogy of free will, whereas in the '*Ode*' lust furnishes a naturalistic metaphor for Tyndall's deterministic materialism, as 'molecules with fierce desires / Shiver in hot embraces', lines that, as was noted earlier, hark back to the apparently affable account of the Norwich address that Maxwell gives in his Liverpool address:

> I have been carried by the penetrating insight and forcible expression of Dr. Tyndall into that sanctuary of minuteness and of power where the molecules obey the laws of their existence, clash together in fierce collision, or grapple in yet more fierce embrace, building up in secret the forms of visible things'.[25]

Maxwell warns against such reductionism, in which self-organizing molecules constitute the germs of complex Darwinian forms of life, in his *Britannica* entry on the 'Atom', where he argues that 'molecular science . . . forbids the physiologist from imagining that structural details of infinitely small dimensions can furnish an explanation of the infinite variety which exists in the properties and functions of the most minute organisms'.[26]

The fundamental phenomenon of the clinamen, in which the individual atom unexpectedly deviates from its path and so hits another, is suggestively modelled by the simple physical bodies of 'In Memory of Edward Wilson' introduced earlier: 'Gin a body meet a body / Altogether free'. The analogy that Maxwell's parody makes of simple physical dynamics to human sexual relations corresponds to that in the first stanza of 'Molecular Evolution', which quietly observes that Lucretius's account of the successive creation of worlds by the clinamen parallels the account he gives of sexual passion in Book IV:

> Usque adeo cupide in Veneris compagibus haerent,
> membra voluptatis dum vi labefacta liquescunt.
> Tandem ubi se erupit nervis collecta cupido,
> Parva fit ardoris violent! Pausa parumper.
> Inde redit rabies eadem et furor ille revisit,

[. . . so eagerly in Venus' toils
They cling, while melt their very limbs, o'ercome
By violence of delight. But when at length
The gathered passion from their limbs hath burst,
There followeth for a space a little pause
In their impassioned ardor. Then once more
The madness doth return . . .][27]

The second quatrain of Maxwell's poem extends the imagery of the 'fortuitous embraces' by describing the existing order and its rupturing with words that translate key terms from Lucretius's account of sexual passion, *haerent* (to hold fast, cling) and *erupit* (bursting forth):

And though they seem to cling together,
    And form "associations" here,
Yet, soon or late, they burst their tether,
    And through the depths of space career.

As in the Lucretian account of passion, where the lovers strive futilely to fuse into one, the atoms in their crucial 'fortuitous embraces' only 'seem to cling together' and after a climactic bursting separate again, thereby furnishing the conditions for a new clinamen, another bout of embraces and the instauration of a new order. The confidently enunciated ontology of the stanza's first quatrain is undermined, deemed merely apparent here. By contracting Lucretius's parallel accounts of the clinamen in Book I and human sexual relations in Book IV, Maxwell highlights the form of reductionism that he criticizes in his article on the 'Atom' and makes the parodic premise of his poem. The effect of this conflation is to present the poem's cosmology as myth rather than science.

The use of the conditional verb 'seem' in the first line of the poem's second quatrain announces an epistemological shift from the categorical statement of the clinamen in the first quatrain to a relativist perspective in the second. The Lucretian cosmology is now presented as a matter of perception rather than objective fact, its contingent formation the consequence of mental as much as physical '"associations"'. The pivotal verb 'seem' in the first line signals the imminent dissolution of a tenuous order of '"associations"', for 'soon or late, they burst their tether'. The metaphor of the restraining 'tether' indicates that the unity of the '"associations"' referred to in the preceding line is not inherent to the various entities they refer to but arbitrary and imposed from without, held together by an extrinsic organizing principle, such as an interpretive prepossession by which an individual might 'form "associations"'.

By effectively making the Lucretian clinamen an allegory for epistemological relativism the second quatrain of 'Molecular Evolution' builds upon the ambiguous transitional reputation that *De Rerum Natura* had in the late 1860s and early 1870s, as it moved from being regarded primarily as a canonical Latin poem full of eccentric speculations to a prescient work of early science. Maxwell acknowledges the altered reputation of the Lucretian doctrine in an observation he makes about 'the atomic theory' in the Liverpool address, that it is 'a branch of physics which not very long ago would have been considered rather a branch of metaphysics'.[28] With the Metropolitans and North Britons a particularly vocal presence at their dinners, the Red Lions were well aware that this atomistic 'branch of physics' accommodated opposing scientific ontologies, a further, vital, dimension of hermeneutical possibility that the Lucretian account had for them. Maxwell's original audience would accordingly have been receptive to his reading of the Lucretian clinamen as an allegory for proliferent interpretations and creative thought, which the poem will identify with the Lions and their playful banter, which it lightly theorizes and defends through its principle of 'Nonsense'.

The confident ontology of the poem's first quatrain is displaced by the second to yield the epistemological correlate of the Lucretian clinamen, a perceptual psychology in which discrete atoms of sense data and other simple mental intuitions 'seem to cling together' in the mind '[a]nd form "associations"'. As well as giving the word a peculiar prominence, the quotation marks encasing '"associations"' make it into a citation that invokes a fundamental and distinctive doctrine of British empiricist philosophy, that of 'the association of ideas'.[29] Although the phrase was coined by John Locke, the doctrine receives its most influential formulation from David Hume, for whom mental impressions and other simple ideas are like contingent particles that the mind draws together in accordance with principles of resemblance, contiguity, and cause and effect. Indeed Hume provides a precedent for the analogy that Maxwell makes suggestively in the first stanza, and confirms in the second, between a scientific ontology and a perceptual psychology, as he describes the workings of the associationist principles in the mind by analogy with the Newtonian law of gravitation: 'Here is a kind of ATTRACTION, which in the mental world will be found to have as extraordinary effects as in the natural, and to shew itself in as many and as various forms'.[30]

The shifting and ambiguous account of the Lucretian cosmology in the first stanza of 'Molecular Evolution' not only accommodates different readings but encourages them. Although the sexual imagery that emerges in the first stanza is consistent with the generic double entendre of Maxwell's earlier poems for the Red Lions, it need not be seen as merely louche. Rather it offers its audience an interpretive choice, an opportunity to exercise the free will that such imagery is used to explain in Lucretius. The foregrounding of the word '"associations"' serves to facilitate and demonstrate the Humean mental function that it apparently cites, for the various contexts it establishes, including its implied audience of Red Lions, encourage the reader to make various mental associations with such things as the Lucretian clinamen, sexual intimacy and the British Association, as well as Hume's doctrine itself.

III

The reverberant hermeneutical provocations of the first stanza of 'Molecular Evolution', as it gestures toward concerns with cosmological and human relations, science and mythology, ontology and epistemology, are enhanced and extended momentously by the second, which declares that its atomist story is a figure for the transformation that the British Association's members undergo after the day's meeting, as these 'British asses' become 'wild Red Lions':

> So we who sat, oppressed with science,
>   As British asses, wise and grave,
> Are now transformed to wild Red Lions,
>   As round our prey we ramp and rave.
> Thus, by a swift metamorphosis,
>   Wisdom turns wit, and science joke,
> Nonsense is incense to our noses,
>   For when Red Lions speak, they smoke.

The poem's exposition of its doctrine of 'Molecular Evolution' is completed here with the introduction of the second term of its central analogy, the current British Association meeting and the members it is addressed to. Parodying the Tyndallic reductionism that Maxwell warns against in his *Britannica* entry on the 'Atom', the poem offers the Lucretian behaviour of atoms as an explanation for the transformation of 'British asses' into 'Red Lions', a mock instance of evolution that suggests a fable, or perhaps one of Ovid's *Metamorphoses*. Maxwell sketches a developmental scheme that includes the poem's audience as its acme, much as Tyndall does in the Norwich address, where, commenting upon biological evolution, he remarks that it is 'a long way from the Iguanodon and his contemporaries, to the President and Members of the British Association'.[31] With its direct references to the BAAS and the Red Lions, the second stanza fully mobilizes the pivotal word of the first, '"associations"', so that the term focuses the poem's audaciously encompassing analogy between the most radical and the most recent points of 'Molecular Evolution'; the original associations of atoms clinging together as molecules and at the other, complex and contemporary, extreme the social and professional associations to which its audience belongs. It economically parodies the sweeping developmentalist narrative of Tyndall's scientific materialism, neatly contracting and caricaturing the hasty ways in which his physicalist argument moves in the Norwich address from salt molecules through to organic nature and human consciousness.

In 'Molecular Evolution' the word '"associations"', foregrounded artificially by its quotation marks and made the nodal point for various contexts that the early stanzas allude to, encourages an overdetermined interpretation in which a series of discrete meanings are summoned and jostle about in various relations of parallelism and punning. This anarchic dialectic of puns and analogies is expanded into the 'swift metamorphosis' by which 'Wisdom turns wit, and science joke'. While the category of 'science' had for Maxwell an early and abiding association with analogy,[32] its transformation into 'joke' here indicates that meaning is not directly explicated, a function that analogy often fulfils, but conversely turned on its head, ostensibly confused, or conflated in the manner of the pun. Such carnivalesque 'Nonsense' is accordingly likened in the poem to 'incense', which suggests not only the exhilarating and intoxicating communion of laughter, but an evanescent essence akin to the often elusive and teasing hybrid meanings that arise in the mind as it works to synthesize the disparate twin terms that compose both analogies and puns. Indeed, the *pungency* of incense suggests a quibble on the semantic concentration and sharp wit of the pun. Similarly, 'Wisdom', good sense drawn from broader experience and thought, 'turns wit', is distilled, like the fumes from incense, a process of paring down that is echoed by the diction, as the two softened syllables of 'Wisdom' are shortened and pointed into the curtailed form of 'wit'.

The second stanza's transformation of the 'British Asses' into 'wild Red Lions' is an allegory for the 'swift metamorphosis' of sense into nonsense, a fable that points a moral about scientific thought and discovery that the remainder of the poem consolidates. The epistemological analogy that the second stanza makes of the first gives parity to the professional theories presented by the British Asses and the 'Nonsense' of the Red Lions, as each is treated as a complementary form of mental '"associations"'. The transmutation of Asses into Lions accordingly clarifies and develops the epistemological schema of the first stanza, the shift it describes from its categorical account of 'all that being hath' in the first quatrain to its undoing in the second, as the elements of this grand cosmological theory 'burst their tether'. The Red Lions are accordingly identified with the liberation of discrete ideas from the 'tether' by which the British Asses hold them in their theories, which the annual meetings of the BAAS review and canonize. This anarchic gesture is identified later in the second stanza with 'Nonsense', the free mental '"associations"' of a new clinamen that ostensibly contradicts the various relations of good sense belonging to wisdom and science.

Identified in the poem with a cognitive function of combination, and contrasted with 'rigid reason' and 'too, too solid sense', 'Nonsense' appears to spring from a principle of imagination, while the analogy with the Lucretian clinamen entails that the combinations it forms are an exercise in free will. They are clearly antithetical to the mechanistic mental combinations of the 'fancy scientific' that Maxwell describes in his '*Tyndallic Ode*', aggregations of 'mental bricks' that suggest an entrepreneurial collocation of ideas '[f]etch[ed] . . . from every quarter'.[33] In 'Molecular Evolution' Tyndall's earnest scheme is drawn into Maxwell's category of nonsense, a rather equivocal judgement here that nonetheless unites him with all of his peers in the BAAS. Hence the fraternal recognition of him as '[t]hat old red lion' at the close of Maxwell's other 'Molecular Evolution' poem, the 'Song of the Cub'.[34]

The foundational analogy that 'Molecular Evolution' makes with Lucretian physics places the sense embodied by the British Asses and the nonsense of the Red Lions in a relation of teasing mutualism. Amidst the original impeccable cosmic order by which the atoms pass evenly through the void, the clinamen emerges as a subtle gesture of defiance that generates further disorder exponentially. This proliferous disorder, however, is seen to constitute a new order, the cosmos *as we know it*, which is identified with the sense of the British Asses and destined in its turn for revolution as the atoms 'burst their tether / And through the depths of space career'. Liberated from their bonds these particles are identified with the Red Lions, and with the nonsense they produce, random concatenations of ideas analogous to the clinamen that may nonetheless establish a new order of thought. As the third stanza puts it, 'the wise . . . / . . . cull those truths of science [from Nonsense], / Which into thee again they turn', a logic of inversion that suggests the role of the Shakespearean fool, and which interpreted through the Lucretian analogy with sexual behaviour constitutes a cyclical pathology of madness. The correlative nature of sense and nonsense suggests Thomas Kuhn's hypothesis that the history of science is marked by shifts in the dominant paradigms used to describe and explain the world, as a prevailing doctrine is contested and superseded by an incommensurable theory, each of which, the established and the new, asserts the privilege of sense for itself and recognizes the other conversely as nonsense. Maxwell characterizes scientific thought accordingly in the Liverpool address as creative, historical and tentative:

> But the mind of man is not, like Fourier's heated body, continually settling down into an ultimate state of quiet uniformity, the character of which we can already predict; it is rather like a tree, shooting out branches which adapt themselves to the new aspects of the sky towards which they climb, and roots which contort themselves among the strange strata of the earth into which they delve. To us who breathe only the spirit of our own age, and know only the characteristics of contemporary thought, it is as impossible to predict the general tone of the science of the future as it is to anticipate the particular discoveries which it will make. Physical research is continually revealing to us new features of natural processes, and we are thus compelled to search for new forms of thought appropriate to these features.[35]

The radical principle of freedom of association that Maxwell's poem opposes to the Metropolitan's deterministic construals of Lucretian ontology and Humean psychology bears a suggestive affinity with Gilles Deleuze's interpretations of Lucretius and Hume. Deleuze writes that 'Hume's originality – or one of Hume's originalities – comes from the force with which he asserts that *relations are external to their terms*'.[36] This is an idea that troubled Maxwell as a young man. However, while many of his early essays and poems are preoccupied with the worrying possibility that reality could be methodically hallucinated by us (through, for example, an elaborate Kantian faculty psychology), this is perceived in 'Molecular Evolution' as a liberating and creative facility of thought. Writing in a 1972 essay on Hume, Deleuze elaborates upon a principle of freedom that he sees such a principle of mind to facilitate:

> The real empiricist world is thereby laid out for the first time to the fullest: it is a world of exteriority, a world in which thought itself

exists in a fundamental relationship with the Outside, a world in which terms are veritable atoms and relations veritable external passages; a world in which the conjunction 'and' dethrones the interiority of the verb 'is'; a harlequin world of multicolored patterns and non-totalizable fragments where communication takes place through external relations.[37]

In his *Logic of Sense* Deleuze finds in the nonsense works of Lewis Carroll demonstrations of this free thought from the Outside and significantly includes in his appendices to this study an essay on Lucretius, 'Lucretius and the Simulacrum', in which he notably describes the ontology of the clinamen with patterns of imagery he invokes to explain the Humean achievement in the passage cited above:

> *Physis* is not a determination of the One, of Being, or of the Whole. Nature is not collective, but rather distributive, to the extent that the laws of Nature (*foedera naturae*, as opposed to the so-called *foedera fati*) distribute parts which cannot be totalized. Nature is not attributive, but rather conjunctive: it expresses itself through 'and', and not through 'is'. This *and* that – alternations and entwinings, resemblances and differences, attractions and distractions, nuance and abruptness. Nature is Harlequin's cloak, made entirely of solid patches and empty spaces; she is made of plenitude and void, beings and nonbeings, with each one of the two posing itself as unlimited while limiting the other. Being an addition of indivisibles, sometimes similar and sometimes different, Nature is indeed a sum, but not a whole. With Epicurus and Lucretius the real noble acts of philosophical pluralism begin. We shall find no contradiction between the hymn to Venus-Nature and to the pluralism which was essential to this philosophy of Nature.[38]

The analogy that Maxwell recognizes in his first stanza of 'Molecular Evolution' between the Lucretian accounts of the clinamen and sexual passion trace their respective '"associations"' to a principle of *inclination*, a subtle but irreducible exercise of free will. The teasing libidinous preoccupations and imagery in which Maxwell exercises such free will so playfully in the poems from the early 1870s is a version of naturalism that asserts and champions the principle of plurality; it involves a model of sexual and emotional inclination that contrasts radically with the naturalism of the Metropolitans, the deterministic lust of the Tyndallic atoms. It is the incursion of matter upon the totalizing purity of thought, the basic principle of passion or inclination, which the clinamen allegorizes so neatly, as the rational Newtonian mechanics of the stream of atoms in their Euclidean plane, an order analogous to the idealist Pythagorean cosmology that the young Maxwell was so drawn to, is disrupted subtly but devastatingly by the swerve of the individual atom. It is this principle that in Maxwell, as in Deleuze and Michel Serres[39] after him, facilitates and recognizes plurality,[40] enables free play, the proliferent nonsense of the Red Lions.

Maxwell's poem articulates and orchestrates the oxymoronic nature of the Lucretian clinamen, the instauration of order as disorder, into his principle of Nonsense, which in turn corresponds to Horace's oxymoronic description of Epicurean thought in Ode 1.34 as 'crazy wisdom' ('insanientis . . . sapientiae', l. 2).[41] Maxwell's reflections upon this text in the poem culminate in a model of the 'combinations of ideas, / Nonsense along can wisely form' that, offering another cheeky counterpoint to Tyndall and the Metropolitans, follows the analogy of Laplace's nebular hypothesis:

> Yield, then, ye rules of rigid reason!
>   Dissolve, thou too, too solid sense!
> Melt into nonsense for a season,
>   Then in some nobler form condense.
> Soon, all too soon, the chilly morning,
>   This flow of soul will crystallize,
> Then those who Nonsense now are scorning,
>   May learn, too late, where wisdom lies.

With the release of the 'tether', the yielding of rigid reason and dis-solving of 'too, too solid sense', these mental capacities '[m]elt into nonsense', the gaseous state identified earlier in the poem with incense and cigarette smoke, that allows atoms of thought to move most freely and dynamically. Indeed, with his principle of the 'Demon' Maxwell has vouchsafed the absolute freedom of individual atoms of gas to con-tradict the great overarching physical (indeed cosmological) principle of entropy, break the second law of thermodynamics. In the passage cited earlier from his Liverpool address Maxwell describes the freedom belonging to human thought as precisely such a defiance of the second law; 'the mind of man is not, like Fourier's heated body, continually settling down into an ultimate state of quiet uniformity, the character of which we can already predict'.[42] The idiosyncratic movement of the individual atom that 'Maxwell's Demon' facilitates is, as was observed earlier, the foundational gesture of the clinamen, of matter as a principle that disturbs the immaculate patterns of rational thought and insists upon the ultimate freedom and autonomy of the phenomenal world, both that of objective nature and subjective mental activity. But then, in accordance with the Laplacean analogy, the radical elements of reason and sense that comprise the dissipated dynamism of nonsense 'in some nobler form condense' to furnish a new universe of understanding, a new paradigm of scientific knowledge.

The free play of the Red Lions dinners, the radical thought experiments of nonsense, gives way to 'the chilly morning' of the BAAS meetings in which their nocturnal 'flow of soul' crystallizes through the complementary analytical modes of thought and scientific procedures demanded by the Association, as they are developed in formal papers presented to the meeting. The Red Lions, which (in a quibble on the cliché 'rant and rave') 'ramp and rave', are figured as rampant in the manner of the heraldic lion, a balance to the more formal and stabilizing British Ass, so that together they furnish the armorial bearings for British science. Once the creative discoveries of the Red Lions are developed and presented soberly in 'the chilly morning' of the BAAS meeting, then, 'too late,' 'those who Nonsense now are scorning' will learn the importance of the poem's teasingly oxymoronic accounts of its relations to wisdom and truth, which are reiterated in the poised quibble made on the last word of the poem, as Nonsense is described as the place 'where wisdom lies'.

Notes

1. The address is reprinted as 'Scientific Materialism' in John Tyndall, *Fragments of Science* (New York: D. Appleton and Company, 1892), 2:75–89.
2. *Report of the Thirty-Eighth Meeting of the British Association for the Advancement of Science*, Norwich, 1868 (London: John Murray, 1869), 3.
3. *Report of the Thirty-Eighth Meeting of the BAAS*, 3.
4. William Paley, *Natural Theology or, Evidences of the Existence and Attributes of the Deity, Collected from the Appearances of Nature* (London: Fauldner, 1810), 1–18.
5. *Report of the Thirty-Eighth Meeting of the BAAS*, 2.
6. *Report of the Fortieth Meeting of the British Association for the Advancement of Science*, Liverpool, 1870 (London: John Murray, 1871), 6.
7. This original is lost, but his Cambridge friend W.N. Lawson recalls 'Maxwell coming to me one morning with a copy of verses beginning – "Gin a body meet a body Going through the air," in which he had twisted the well-known song into a description of the laws of impact of solid bodies'. Lewis Campbell and William Garnett, *The Life of James Clerk Maxwell* (London: Macmillan, 1882), 174–75.

8. Campbell and Garnett, *The Life of James Clerk Maxwell*, 630.
9. *Report of the Fortieth Meeting of the BAAS*, 1–2.
10. Campbell and Garnett, *The Life of James Clerk Maxwell*, 634.
11. Gowan Dawson, *Darwin, Literature and Victorian Respectability* (Cambridge: Cambridge University Press, 2007), 106.
12. James Clerk Maxwell, 'Molecular Evolution', *Nature*, 8.205 (2 October, 1873), 473; Campbell and Garnett, *The Life of James Clerk Maxwell*, 637–38.
13. See Herbert Spencer, *Principles of Psychology* (London: Longman, Brown, Green and Longmans, 1855), *First Principles* (London: Williams and Norgate, 1862), *Principles of Biology*, 2 vols (London: Williams and Norgate, 1864–67), *Principles of Psychology*, 2nd edition, 2 vols (London: William and Norgate, 1870–72).
14. Campbell and Garnett, *The Life of James Clerk Maxwell*, 637.
15. Frank M. Turner, *Contesting Cultural Authority: Essays in Victorian Intellectual Life* (Cambridge: Cambridge University Press, 1993), 262–83.
16. Fleeming Jenkin, 'The Atomic Theory of Lucretius', *North British Review*, 48 (1868), 211.
17. See, for example, Maxwell's early essay 'On the properties of matter', in P.M. Harman (ed.), *The Scientific Letters and Papers of James Clerk Maxwell* (Cambridge: Cambridge University Press, 1990–2002), 1:113.
18. Jenkin, 'The Atomic Theory of Lucretius', 220, 222.
19. Fleeming Jenkin to James Clerk Maxwell, 28 October 1871, Cambridge University Library, Add 7655/II/51.
20. Harman, *The Scientific Letters and Papers of James Clerk Maxwell*, 2:250.
21. James Clerk Maxwell, 'On the Dynamical Theory of Gases', in W.D. Niven (ed.), *The Scientific Papers of James Clerk Maxwell* (Cambridge: Cambridge University Press, 1890), 2:27–28.
22. Lucretius, *De Rerum Natura*, 3rd edition, trans. H.A.J. Munro (Cambridge: Deighton Bell, 1886), Bk. II, lines 276–80, 283–96.
23. James Clerk Maxwell, 'Discourse on Molecules', in W.D. Niven (ed.), *The Scientific Papers of James Clerk Maxwell* (Cambridge: Cambridge University Press, 1890), 2:373.
24. M. Tullius Cicero, *Of the Nature of the Gods* (London: Franklin R., 1741), I.xxiv.
25. *Report of the Fortieth Meeting of the BAAS*, 1–2.
26. James Clerk Maxwell, *Britannica* entry on the 'Atom', in W.D. Niven (ed), *The Scientific Papers of James Clerk Maxwell* (Cambridge: Cambridge University Press, 1890), 2:461.
27. Lucretius, *De Rerum Natura*, trans. Charles E. Bennett (Roslyn, NY: Walter J. Black, 1946), Bk IV: lines 13–19, 220.
28. *Report of the Fortieth Meeting of the BAAS*, 4.
29. John Locke, *An Essay Concerning Human Understanding*, ed. Peter H. Nidditch (Oxford: Clarendon, 1975), Bk. II, 394–401.
30. David Hume, *A Treatise of Human Nature*, ed. L.A. Selby-Bigge (Oxford: Clarendon, 1978), Bk. I, Part I.V: 12–13.
31. *Report of Thirty-Eighth Meeting of the BAAS*, 6.
32. See his 1856 Apostles essay, 'Are There Real Analogies in Nature?,' in P.M. Harman (ed.), *The Scientific Letters and Papers of James Clerk Maxwell* (Cambridge: Cambridge University Press, 1990–2001), 1:376–83.

33. Campbell and Garnett, *The Life of James Clerk Maxwell*, 636.
34. Campbell and Garnett, *The Life of James Clerk Maxwell*, 638.
35. *Report of the Fortieth Meeting of the BAAS*, 8.
36. Gilles Deleuze, *Pure Immanence: Essays on a Life*, trans. Anne Boyman (New York: Zone, 2001), 37.
37. Deleuze, *Pure Immanence*, 38.
38. Gilles Deleuze, *The Logic of Sense*, trans. Mark Lester (London: Athlone, 1990), 267.
39. Michel Serres, *The Birth of Physics*, trans. Jack Hawkes (Manchester: Clinamen Press, 2000).
40. Cf. Deleuze, *Logic of Sense*, 279.
41. David West, *Horace Odes I, Carpe Diem: Text, Translation and Commentary* (Oxford: Clarendon Press, 1995), 161.
42. *Report of the Fortieth Meeting of the BAAS*, 8.

8
Writing the 'Great Proteus of Disease': Influenza, Informatics and the Body in the Late Nineteenth Century

James Mussell

When I began to draft this chapter, in April 2009, a computer worm called Conficker began to download information from the internet onto the machines that it had infected. Worms take advantage of information transfers to move between machines and are capable of replicating themselves from within the systems they infect. Although not harmful in themselves, they are often used to distribute malicious code that alters the performance of an infected machine.[1] As soon as Conficker began downloading, information spread over the micro-blogging site Twitter (http://twitter.com). Various news sources around the world, both specialist technology sites and traditional news vendors such as the BBC, broke the news on Twitter, and their followers promptly reproduced this with short messages ('tweets') of their own. Within four days, however, discussion on Twitter was dominated by a different kind of virus, as news of the first death caused by an outbreak of swine flu was reported in Mexico. In each case, Twitter distributed news just as other bodies, those of computers and people, propagated viruses of their own.

In this chapter I argue that the influenza pandemics of the late nineteenth century were simultaneously biological and cultural phenomena. However, I do not mean that the impact of the biological virus caused a range of cultural effects or even that the diverse cultural effects of the virus prompted it to appear in a variety of material forms. Rather, because of its indeterminate material nature influenza was able to infect both the human and the social body simultaneously and move between both. News, computer worms and biological viruses all exploit informational networks but none are purely information. Each has a material supplement, whether it is language, computer code, electrical signals or chemical compounds, that dictates how they move information through culture. Just as the behaviour of Conficker and the messages on Twitter are determined by the way in which each is coded and executed, so the ontological status of biological viruses such as influenza limits their transmission and effect. You cannot catch the flu by reading about it and the possibility of a computer virus affecting biological systems remains a fantasy of science fiction.[2] Yet information, essentially immaterial yet dependent upon material media and the systems that connect them, operates on and across the borders of minds, bodies and machines. Although conceptualized as independent of the media upon which it relies, information does not exist in the abstract, only as part of an object in some sort of system of communication.[3] As information, by definition, is capable of transmission, of transcending its material manifestation, the extent to which the news, computer worms and biological viruses differ from being purely information is the extent to which they can be attributed bodies of their own.

When influenza appeared in 1889 it was expected that it too would be attributed a body. Robert Koch's announcement of the *Tubercle bacillus* in 1882 not only demonstrated that germs were the causative agents for disease but that it was also possible to isolate and identify them.[4] Indeed, Koch's four postulates, which still provide the criteria for establishing a causal link between a microbe and a disease, were predicated upon the idea that diseases are caused by discrete entities that could be isolated from and grown separately to diseased tissue.[5] Without a demonstrable causative agent for influenza, it was difficult to cohere the diverse symptoms with which it was associated to a single named condition and impossible to delimit its modes of transmission and contagion. Highly contagious but seldom fatal, and causing a range of symptoms that were identical to those of other conditions, influenza could pass between bodies and also enter into them. All viruses, whether electronic or biological, challenge the autonomy of individual bodies by foregrounding both their interconnectedness and their porousness: however, the indeterminate materiality of the late nineteenth-century influenza virus permitted it to exploit a range of media, whether human or nonhuman, and so infect biological and informational systems simultaneously. Although reading about the flu was not the same as having the flu, it appeared that it could be spread by communication networks as well as human contact. Like a computer virus, influenza utilized the networks that connected disparate bodies and exploited human activity to reproduce: however, as an infection of information, it was not restricted to electronic systems, but instead became as immaterial as information itself.

* * *

The electronic communication of information, whether human-readable or machine-readable, is considered to be one of the defining characteristics of the information age. The speed with which electronic networks function permits humans and machines distributed across the globe to interact almost instantly. This fantasy of proximity has been fostered by the development of web 2.0 technologies that not only make it easier to create content for distribution online, but also to liberate information from the applications with which it was created. Although the popularity of web 2.0 resources such as Flickr (www.flickr. com), Facebook (www.facebook.com) and Twitter is due to the connections they facilitate between users, their power lies in the way they aggregate and order the information that users provide (often for free). By recording the behaviour of their users over time, such resources offer back an alienated, disembodied and fractured version of the self that differs from postmodern subjectivity in that it lacks any material repository. When allied with the more familiar effects of the information age – instant communication, electronic financial transactions, digitization and digital simulacra – it seems that it is not just the signified that is deferred, but materiality itself.

This has lead some to overdetermine the boundary between the dematerialized present and the material past. In his *Living on Thin Air* (1999), Charles Leadbeater set out the transformations that he believed were necessary for societies to thrive in the electronic age:

> We have to go forward because if we retreat we end up with gridlock. Our societies and governments often seem paralyzed, or at best enfeebled, in the face of economic and technological change that outstrips their capacity to respond. We are weighed down by institutions, laws and cultures largely inherited from the industrial nineteenth century; yet we confront a global economy driven by an accelerating flow of new ideas and technologies which are creating the industries and products of the twenty-first century. We have welfare systems which are impervious to reform, parliamentary systems which are recognizably Victorian and schools which still resemble their nineteenth-century forebears. Imagine fighting a modern war using cavalry: that is the position we are in.[6]

The Victorian here stands for an industrialized past populated with things in contrast with a present confronted with the rapid electronic transfer of information. The association of materiality with history corresponds to a blinkered version of globalized culture that is blind to the production and circulation of commodities. The postindustrial West is presented as the location of (post)modernity, while the industrial activity that sustains it is pushed to the margins and associated with a less civilized past.

The emphasis on the immateriality of information at the expense of the materiality of informatics also elides the mechanism through which the past influences the present. Existence through time is always predicated on the material. Our inheritance is always embodied, whether this is in terms of genetics or property, and the Victorian institutions that Leadbeater condemns as anachronistic survive precisely because they are structures that are intended to embody memory.[7] What this material inheritance demonstrates, in diverse contexts ranging from spiritualism to science and aesthetics, is that the Victorians, although an industrialized society, were not only concerned with the immaterial, but also its relation to the material world.[8] The resurgence of interest in nineteenth-century material culture has gone some way to restoring the role of objects in culture but often stresses their cultural meanings, particularly as commodities, at the expense of their material properties.[9] The two, of course, are never wholly divorced as material form has semiotic potential but, as Bill Brown has noted, materiality, the 'thingness of objects', only becomes apparent when 'they stop working for us'.[10] The edges of things both account for the integrity of objects, allowing them to exist in the world, and their resistance, as they are literally what return our touch. To become carriers of information, the edges of objects must be overcome so that they function as nodes in networks rather than things. Objects in informational networks thus transcend their boundaries in order to influence one another. In this way, information plays a part in defining the limits of objects and determining what meaning they have for us. Not only does information help determine the properties of its material media, but it also hints at other objects of which we might not be fully aware.

The etymology of the word 'virus' captures its material ambiguity. Joost Van Loon argues that the 'virus has always functioned as a label for that which cannot be named otherwise, a remainder of the known world, and a reminder of nature's inherent unintelligibility'.[11] The word 'virus' carries with it an imponderable supplementarity, similar to the way 'thing' describes that aspect of materiality that always lies beyond the human world.[12] Derived from a Latin word meaning 'slimy liquid' – a definition that lived on in English as an obsolescent term for semen – the meanings of 'virus' prior to the emergence of its contemporary biological definition were all associated with poisonous discharges.[13] As harmful substances, viruses are conceived as being alien to the body, but their malicious influence, affecting the body from within, makes their boundaries difficult to establish. Just as Brown's description of things foregrounds their liminal status, hovering 'over the threshold between the nameable and unnameable, the figurable and unfigurable, the identifiable and unidentifiable', so the virus both names something that can be transmitted but at the same time is yet to be defined.[14] The body's immune system must identify viruses as other, marking them so that they can then be destroyed. This is usually understood in informational terms, with chemical signals establishing the boundaries between self and other within an otherwise self-contained system; but what is destroyed is, nonetheless, an object.[15] The same process underpins the detection and destruction of computer viruses. Like biological viruses, they too are parasitic – pieces of alien code that insert themselves into an application so that they can be executed as part of the system's legitimate processes – and virus protection software works by checking code against indices of either 'healthy' programs or known threats in order to establish what to destroy. As Stefan Helmreich argues, although protection against computer viruses is frequently understood in biological terms (my virus checker puts suspected viruses into quarantine), the extension of biological language to electronic systems exposes its ideological foundations. In an age of networked computing, it is not clear what constitutes the body that needs protecting: is it the workstation, the network, or the web as a whole? Although immunology might fall back on the integrity of the body, human bodies, of course, are also networked.[16] The immune system establishes the difference between self and other, but it does so from within the body, suggesting that not only are our bodies connected, but that they also overlap. Information – as language, pattern or difference – might be the means in which bodies overcome their boundaries and influence one another, but it is also the means through which immune systems establish what those edges and so what those objects are.

When the outbreak of influenza in St Petersburg was announced in *The Times* on 25 November 1889, the paper's correspondent linked it to the sanitary conditions of the city and Russia as a whole.[17] However, by 3 December the rapid distribution of influenza around the Russian Empire prompted predictions of its spread into the rest of Europe.[18]

The previous documented outbreak in Britain had been in 1847 and the fullest account of the disease was Theophilus Thompson's *Annals of Influenza* from 1852, which provided a natural history of the disease from 1510 to 1837. Despite its high morbidity – early reports stated up to a third of the population of St Petersburg had been afflicted – the low number of deaths resulting from influenza reassured British commentators that its effects, should it spread, would not be serious. In fact, *The Times* suggested that 'in the interests of science, its arrest in Russia would probably be a misfortune; for, if it should reach more scientific countries, it will certainly be subjected to a more searching examination than is otherwise at all probable'.[19] The improved sanitary conditions that had prevented the spread of cholera from across the channel during the recent outbreak in France would also protect the population from the more severe effects of influenza. In addition, confidence in contemporary microbiology meant that the outbreak offered the opportunity to learn more about the germ that was suspected to be its cause. As *The Times* stated:

> To the scientific pathologist, indeed, the reappearance of a disease which has not visited this country for forty-two years would be a matter of the most lively interest calculated to excite in him emotions parallel to those with which a sportsman would engage in the pursuit of some previously unknown game, or an archaeologist, before mummies were vulgarized, in the task of unrolling one.[20]

Its reappearance offered the opportunity to transform the symptoms of influenza from occulting surface phenomena into demonstrable marks of presence. Like a spirit summoned for a séance, influenza would be made to perform subject to scientific scrutiny so that it could be allocated a place in the known material universe.

Influenza reached Britain late in December 1889. It is difficult to give a precise date because it was not clear when influenza actually became manifest. The word was already in use to describe a range of catarrhal conditions and, particularly in winter, there were a range of drugs advertised that promised to cure it. However, once the severity of the outbreak in Russia began to be appreciated, publications began to take news of its spread from the news agencies and their own correspondents around Europe. Reuters, for instance, ran an 'Influenza Special' that supplied updates on the progress of the pandemic to the British press. London was one of the last capitals to experience influenza, so most people's knowledge of the outbreak was limited to such reports.

The Times was the first British paper to publish news of the outbreak in St Petersburg on 25 November 1889 and it continued to print reports of influenza from Reuters and its own correspondents.[21] By 30 November it was reported to have spread around Russia;[22] by 9 December it was in Berlin[23] and Copenhagen,[24] and by 10 December Reuters reported it was also in Austro-Hungary.[25] By 11 December reports appeared confirming cases in Paris.[26] On 13 December, *The Times* dismissed rumours that it had broken out in Chiswick, West London, as 'fictitious terrors',[27] but the following week reported outbreaks as far apart as Madrid,[28] Belgrade,[29] Amsterdam and New York.[30] Readers in Britain, therefore, had five weeks to trace the progress of the pandemic from city to city, but did so with the creeping awareness that influenza travelled along the same routes and utilized the same technologies as the news they were reading.

When influenza became imminent in London, the press found itself in its usual position of warning about the dangers of rumour while being responsible for peddling it.[31] The Prime Minister, Lord Salisbury, was one of the first British victims of the pandemic. Answering the question of whether he actually had influenza, the *Pall Mall Gazette*, in its gossip column, 'To-Day's Tittle Tattle' replied:

> That is the question on everyone's lips for the next day or two. As nobody seems to know exactly what the influenza is, it must be still more difficult to say whether any particular patient has it or not; and there is thus a boundless field for speculation and gossip.[32]

Lori Loeb has shown how influenza challenged the authority of the medical profession.[33] Unable to pronounce definitively as to what influenza was and reduced to suggesting treatments that might ameliorate its symptoms rather than provide a cure, medicine could offer little more than the commercial market for drugs.[34] The press was well placed to benefit from this situation, able to sell news of the pandemic, disseminate medical advice, and advertise a wide range of drugs. The *Pall Mall Gazette*, for instance, used a 'lady reporter' to pose as a patient in order to obtain medical advice from four leading practitioners. Although the paper justified the publication of this advice on the basis that the public needed information, it scandalized the medical profession as it not only revealed that their prescriptions were all slightly different and generally confined to treating the symptoms rather than the germ itself, but also destroyed the market for their expertise by providing access for the price of a newspaper.[35] The manufacturers of drugs, on the other hand, were well situated to capitalize on rumours of the approaching pandemic and soon began to tailor their advertisements accordingly. As drug manufacturers did not need to reveal their ingredients unless they contained an ingredient controlled by the 1870 Pharmacy Act, they encouraged the spread of information about both the disease and their drugs. On 4 January 1890, for instance, shortly after the first cases were confirmed in London, an advertisement for Salt Regal in the *Illustrated London News* claimed, under a headline of 'THE COMING EPIDEMIC! THE COMING EPIDEMIC!!', that it was a 'Preventive and Safeguard'.[36] Salt Regal was advertised in a wide range of publications, including the *Pall Mall Gazette*, the liberal weekly *Truth* and the evangelical weekly *Great Thoughts*. At the peak of the pandemic, its manufacturers ran a subsequent advertisement that claimed the 'users of SALT REGAL have hitherto escaped the EPIDEMIC', reminding readers once more that 'SALT REGAL [was] a Preventive and Safeguard!!' and offered a range of testimonies as evidence.[37]

The rapid diffusion of information about influenza – whether news, gossip, commentary or advertisement – was difficult to separate from its spread through the bodies of those infected. Its high morbidity coupled with its low mortality meant that experience of the flu was widespread while very few people died from the condition. There were four waves of influenza between 1889 and 1894 (January–February 1890; April–May 1891; January–February 1892; December–January 1893–4) and it is estimated that a third of the adult population of England, Wales and Ireland suffered at least one attack over these years.[38] There were further outbreaks in 1895 and 1899–1900, but all these late nineteenth-century occurrences of influenza tend to be overshadowed by the 1918 outbreak that was responsible for an estimated 20 million deaths world-wide.[39] Unlike the 1918 outbreak, with its high rate of mortality, these earlier pandemics were characterized by a low mortality rate and most deaths were attributed to associated conditions such as pneumonia. For instance, during the first outbreak in 1889–90, only 599 deaths were attributed directly to influenza in London, rising to a maximum of 2205 during the more severe outbreak in April–May 1891. Even when the other deaths in which influenza was thought to have contributed are added, these totals reach only 2800 and 5800 respectively.[40] In a city of around 5 million of whom up to 3 million caught the disease, the probability of dying from a bout of influenza was very low.[41] Unlike cholera, which had a high mortality rate regardless of the age of those afflicted, influenza tended to be more dangerous for the elderly, the poor, or those susceptible to other respiratory diseases and so was not taken quite so seriously as a threat to public health. That said, the more lethal second wave did prompt some concern. In the *Spectator*, for instance, fears were expressed for public order as government was by then 'comparatively old' and, in *Leisure Hour*, Alfred Schofield reminded his readers that even though influenza had a low mortality rate (he cited 1 per cent) its high morbidity meant that it had still killed more than cholera.[42] The overall consensus, however, was that influenza was something that simply had to be tolerated.

Prior to the outbreak in Britain, the *Lancet* had noted that its rapid spread meant that it had 'no geographical limitation, it is apparently uninfluenced by season or climate, and its virus travels over sea and land in a manner so baffling and contradictory to the ordinary conceptions of the transmission of infection as to render any simple explanation of its nature almost impossible'.[43] As the symptoms of influenza resembled so many other illnesses it was difficult to determine whether it was present in a population until it was sufficiently widespread and then, of course, it was too late to study its propagation. Equally, once influenza was mooted as a cause of illness within a population, there was concern that symptoms caused by other diseases would be misattributed to it. The only sign of its presence was its rapid communication. An attack of influenza typically lasted two to four days, with a period of recovery lasting another two weeks. As each wave of the pandemic lasted approximately four to six weeks, with the number of cases steeply declining towards the end of this period, the bulk of cases occurred simultaneously.[44] No biological organism, it was believed, could travel this fast and, despite various suggestions as to the transmission of either the germs of influenza or some sort of force to activate them, many accepted the outbreak as another manifestation of late nineteenth-century mass culture.[45] Influenza rapidly became known as the 'fashionable epidemic' and there were suspicions that some complaining of the influenza were either exaggerating the symptoms or faking them entirely.[46] For instance, the *Illustrated London News* ran a full-page cartoon that depicted a number of people either taking advantage of the influenza to indulge or blaming it for the consequences of overindulgence. Andrew Wilson, who contributed its science column entitled 'Science Jottings', grudgingly devoted his 'Monthly Look Around' to the raging pandemic on the basis that from 'peer to peasant everyone appears to regard this topic as the only subject of interest'. Wilson, a lecturer on zoology and comparative anatomy at

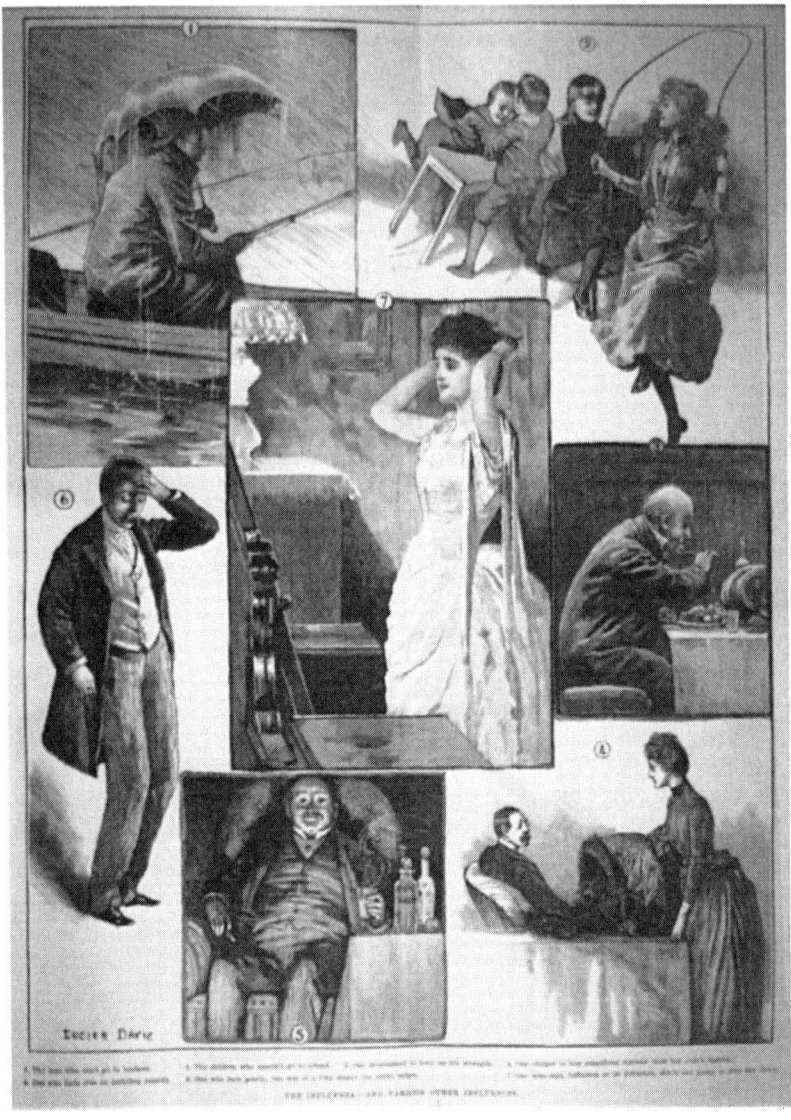

Figure 9: 'The Influenza: And Various Other Influences', *Illustrated London News*, 95 (18 January 1890)

the Edinburgh Medical School, had little to tell his readers about the scientific underpinnings of the pandemic:

> The science of influenza is summed up by saying that doubtless it is a germ-produced disease. Only on the theory that its germs were diffused far and wide can we account for its spread.[47]

Instead, he speculated that influenza has been 'used as an excuse for the breaking of unwelcome engagements' and that 'the next comic song of the music-halls will be founded on the epidemic, with a chorus beginning "Have you got it?" or some equally asinine refrain'.[48] Rather than give the microbiological context for the pandemic and ignoring the (false) announcement that the germ had been discovered in Vienna, Wilson instead attacked the medical profession's concerns about self-doctoring, which he termed 'old wifeism in medicine'.[49]

It was to 'The Ladies' Column' of the *Illustrated London News* that one turned to receive useful information about influenza and advice on how to treat it. In the same issue as Wilson's remarks, Florence Fenwick Miller stated that influenza had 'proved not so trifling and half-imaginary a complaint as was at first hoped' and then gave statistical information as to its effect on the death rate in London.[50] She claimed that doctors had 'always been clever at concealing their helplessness against disease under a cloud of imposing language' before explaining the terms 'microbe' and 'bacillus':

> It is a somewhat appalling idea that each human system forms a world, in which a whole myriad of microscopic animalculae are born, live by their own exertion, perhaps form kingdoms or republics, hoard wealth, prey on each other, rear offspring, and depart from life after what seems to them a prolonged existence unaware that the microcosm in which they have passed their span is not the universe! But though there is an awful poetry in this reflection, it does not help us to a knowledge of how to keep out, or how to exterminate after having let in, those microbes which are inimicable to the body's wellbeing.[51]

Miller's evocation of the microbiological sublime emphasized the porousness of the body. For Miller, it was medicine's responsibility to intercede into the subvisible realm and provide the immunological system that would protect the autonomy of the individual, but it was the responsibility of her readers ('us housewives') 'to tackle the difficulty, so to speak, from the opposite side, and take care that our patients are given all that they can take to maintain the general strength to fight against the deductions made from it by the microbe army'. Rather than target the microbe, Miller suggested nourishing the body, reasserting its outside, and so substituted effective nursing for ineffective doctoring.[52]

For Miller, the microbiological sublime offered a way of imagining the body in the world that did not depend on its visible limits, but she nevertheless argued that modernity, particularly information technology, had put these limits at risk. She imagined that 'the enterprising microbe has travelled with speed, perhaps by train or by telegraph, to other parts of the kingdom' and warned that 'we must pay the price of our advantages, and the trains which carry us and our letters so rapidly about serve also, apparently, for the quick conveyance of the destructive microbes of disease'.[53] What is interesting is that, despite her evocation of influenza as a microbe, it here travelled like information, like letters on trains or electrical signals over wires. In the body too, influenza seemed to move rapidly, affecting different parts simultaneously and preventing the symptoms from appearing as indexical marks that could delineate the hidden microbe. Sir Morrell McKenzie, one of the physicians caught by the *Pall Mall Gazette*'s undercover reporter, attempted to account for the action of influenza in the *Fortnightly Review* shortly after the second wave in 1891. He explained that 'influenza is the very Proteus of disease, a malady which assumes so many different forms that it seems to be not one, but all diseases' epitome, and its symptomatology includes almost everything, from running at the nose to inflammation of the brain'. He insisted that the cause of the disease was 'a living germ of some kind' but, in order to account for its multiple effects, suggested that it had an 'electric affinity for the nervous system'.[54] Rather than attack parts of the body, influenza instead exploited the means through which they were connected. McKenzie compared the effect of a bout of influenza to an electrical storm, explaining that the 'extraordinary disturbance in our telegraphic system sometimes caused by a thunderstorm is as nothing compared with the freaks played by the living conductors in the human body if anything throws the governing centres out of gear'.[55] Affinity expressed a structural relationship as well as an attraction and the analogy between the nerves and telegraph was an old one. In suggesting that influenza might act upon the nerves, McKenzie tantalizingly hinted that it might also move through them.

Neither the social body nor the human body provided the friction of materiality for influenza, and so it behaved like an informational flow rather than the microbe it was widely suspected to be. As signal, rather than thing, its effects signified its presence but could not bring the microbe itself into being. Its edges, in other words, eluded the various technologies that attempted to make it an object of discourse. Not only was it difficult to establish causal relationships between the effects of influenza and whatever caused it, but there was no 'it' for such indexical marks to point towards. Competing discourses attempted to provide contexts within which influenza could cohere, but the result was multiplication and ethereality rather than a material object articulated through chains of signification. N. Katherine Hayles distinguishes between embodiment and the body in order to account for the tendency to dematerialize the body in postmodern thought. For Hayles embodiment is 'contextual, enwebbed within specificities of place, time, physiology and culture that together comprise enactment', whereas the body is the product of discourse, an idealized form that is 'always normative relative to some set of criteria'.[56] In the case of the late nineteenth-century influenza pandemic, there were many embodied experiences of influenza but there was not a coherent, unified, bounded construction of the body that could provide the discursive terrain within which the microbe might substantiate itself. Not only did the virus move through and between human bodies, but it also seemed to move with the news, on trains, through the post or over the wires. Rather than simply manifest itself in the human body, its disparate effects situated influenza across a variety of discursive terrains simultaneously. For Hayles, the tendency to reduce the body to informational flows in postmodern thought – to become posthuman – is due to the neglect of the specific instantiations of embodiment. However, for influenza the reverse was true. As its symptoms were not confined to the recognized edges of human bodies and its spread not restricted to the recognized ways in which such bodies were connected, it could not materialize as an objective entity independent of its individual instantiations. Instead, it became like information, able to move through what appeared to be different bodies and so reveal the hitherto concealed connections between them.

*　*　*

A week before the conference that prompted this collection of essays there was a terrorist attack at Glasgow airport. Two men drove into the airport terminal, crashing their vehicle and setting it on fire. In a news report that followed the incident, a newsreader for Channel 4 News stood in front of footage of the still-burning car and explained that it remained a crime scene as it could provide forensic evidence despite the evident damage. In the same report, reference was made to hard disks and mobile phones that had been obtained from the suspects' homes. These objects, we were told, continue to carry records of communications somewhere within them even though the data might have been deleted. To reinforce his point, the newsreader reminded his audience that during the investigation into the murders in Soham in 2002, police obtained a breakthrough when the accused was confronted by evidence that one of the victims sent a text message (SMS) from near his house. In all these cases information was not conceived as immaterial and transitory, passing through the media that it employed, but instead as something that left traces that could be recovered. Just as material media create noise, affecting the signal, so the signal affects the media, depositing its traces.

Information is never encountered outside of objects and it is the objects that provide the discursive contexts within which information must be read. In terms of forensics, it is the task of the scientist not only to extract information, but to identify the correct information, leaving the rest as part of the object. Reading, too, is a process which differentiates between the code that is to be deciphered and all the other meaningful aspects to a text. In each case an interpretive process decides what remains as part of the object and what should be liberated from it. Just as the immune system reacts to identify and objectify the thing that threatens it, so too do institutions, whether this is medical science, trying to define the microbial nature of influenza by restricting it to the body, or the State using forensic science to identify which of its citizens represent a threat to its security. Such institutions exist to establish limits to the agency of people and things, and they do so by attributing influences to discrete entities.

Influenza, as various nineteenth-century commentators noted, is etymologically linked to the idea of influence. *The Times*, for instance, told its readers that the word originated in Italy and was connected to older traditions that associated instances of the plague with the movements of astral bodies.[57] Late nineteenth-century space was not empty but provided the medium through which things – and indeed people – could influence one another. The concept of the ether, and the electromagnetic principles for which it furnished an explanation, provided a mechanism for discrete objects to overcome their material edges and affect, through emissions, vibrations and fields, other bodies around them. The various social fads and fashions, advertised and promoted in the press, that swept through the newly constituted middle-class masses, demonstrated how people could be enthralled by products as well as by political or social movements.[58] The evangelical weekly *Great Thoughts*, for instance, reminded its readers that influence was not, in itself, good or bad, but was simply made more pronounced by modern technology. 'There is far more intercourse and sympathy between men now than in the distant past,' the anonymous author warned, 'whether we will or not, we are always affecting those around us, always transmitting good or evil'.[59] Just as it was up to the individual to modulate their influence so as to benefit others, it was also for the individual to determine which influences should be accepted and which should be rejected. However, it was not always easy to distinguish between good and bad influences, or even to determine where the self ended and influence began. The same technologies that exposed the subject to influence also distributed subjectivity much in the same way as telepathy permitted contact between minds.[60] Influences could be seductive and individuals were weak: the trial of Oscar Wilde testified to the virulent influences that underpinned decadent aesthetics, while also pruriently guarding against their further transmission.[61] Even within the self, subjectivity was understood as the performance of boundaries: for Freud consciousness was the result of the repression of influences; for Myers it was simply those influences of which we were aware.

Harold Bloom, in his *Anxiety of Influence*, claimed that 'Influence is *Influenza* – an astral disease. If influence were health, who could write a poem? Health is stasis'.[62] The unhealthiness of influence tells of its disregard for boundaries; yet, in acknowledging influences as other, as outside, or as belonging to a particular entity, we also identify the boundaries that they transgress. For instance, influence both challenges the idea of the subject as rational and self-determining while also, through the notion of repression, providing the mechanism for establishing its boundaries and its continuity. What is important to note is the reflexivity of this process: influences are only detectable at the point of contact between entities; one might influence the other, but it is the other's resistance that defines its contours. Objectifying something establishes its edges, but this applies to whatever is doing the objectifying as well.

The various effects of influenza, distributed across and between disparate discursive bodies, could not function as signs in an economy of presence and absence.[63] They indicated that influenza was present, but could not reify this presence into an object with edges of its own. Although the dominant explanation for its effects was some sort of microbe, the experience of witnessing its spread made it appear like an early computer virus, infecting informational systems so that they spread it further. Computers are vulnerable to viruses as they run executable programs, often delivered over networks. However, these programs are not pure information but are encoded electronic signals and, to defend against them, software discriminates between those influences that are permitted and those that must be excluded. As the late nineteenth-century influenza manifested itself in diffuse and often exclusive discursive terrains, it was difficult to delimit the edges of its influence or restrict it to a particular material environment. The use of the term 'viral' to describe the transmission of something through culture is a relatively recent addition to the *OED*, first appearing in 1989 with reference to viral marketing. The nineteenth-century influenza pandemic demonstrates that what makes viruses viral is their indeterminable materiality. By failing to delimit its influences and so reify its cause, the flu was able to exploit its interfaces within and between bodies in order to mutate from a thing to a message and pass virulently through culture.

Notes

1. Eugene H. Spafford, 'Computer Viruses as Artificial Life', *Journal of Artificial Life*, 1.3 (1994), 249–65.
2. See, for instance, Neal Stephenson's novel *Snow Crash* in which a narcotic in a virtual world called the metaverse distributes a virus that affects both the virtual avatars and the people who control them. My thanks to Jay Clayton who alerted me to this classic cyberpunk novel.
3. N. Katherine Hayles, 'Virtual Bodies and Flickering Signifiers', in Timothy Druckery (ed.), *Electronic Culture* (New York: Aperture, 1996), 259–60.
4. Thomas D. Brock, *Robert Koch: A Life in Medicine and Bacteriology* (Madison, WI: SciTech Press, 1988), 117–39.
5. Brock, *Robert Koch*, 179–82.
6. Charles Leadbeater, *Living on Thin Air: The New Economy* (London: Penguin, 1999), 23.
7. Bruno Latour, *Pandora's Hope* (Cambridge, MA: Harvard University Press, 1999), 151–56.
8. Marina Warner, *Phantasmagoria: Spirit Visions, Metaphors and Visual Media into the Twenty-First Century* (Oxford: Oxford University Press, 2006).
9. See Thomas Richards, *The Commodity Culture of Victorian England: Advertising and Spectacle, 1851–1914* (London: Verso, 1991); Asa Briggs, *Victorian Things*, 3rd edition (Stroud: Sutton, 2003); Elaine Freedgood, *The Ideas in Things: Fugitive Meanings in the Victorian Novel* (Chicago: University of Chicago Press, 2006); Clare Pettitt, 'On Stuff', *19: Interdisciplinary Studies in the Long Nineteenth Century*, 6 (2008), 1–12.
10. Bill Brown, 'Thing Theory', *Critical Inquiry*, 28 (2001), 4.
11. Joost Van Loon, 'A Contagious Living Fluid: Objectivity and Assemblage in the History of Virology', *Theory, Culture and Society*, 19.5–6 (2002), 108.
12. Brown, 'Thing Theory', 4–5.

13. See the *Oxford English Dictionary*; also Patrick Collard, *The Development of Microbiology* (Cambridge: Cambridge University Press, 1976), 159–60; and Van Loon, 'A Contagious Living Fluid', 108.
14. Brown, 'Thing Theory', 4–5.
15. Emily Martin, 'The End of the Body', *American Ethnologist*, 19.1 (1992), 123–24.
16. Stefan Helmreich, 'Flexible Infections: Computer Viruses, Human Bodies, Nation-States, Evolutionary Capitalism', *Science, Technology and Human Values*, 25.4 (2000), 474–79.
17. 'Russia', *The Times* (25 November, 1889), 6.
18. 'The Influenza Epidemic in Russia', *Lancet*, 134 (1889), 1134; and 'London: Tuesday, December 3, 1889', *The Times* (3 December, 1889), 9.
19. 'London: Tuesday, December 3, 1889', 9.
20. 'London: Tuesday, December 3, 1889', 9.
21. 'Russia', 6.
22. 'Russia', *The Times* (30 November, 1889), 5.
23. 'Germany', *The Times* (9 December 1889), 6; and 'The Influenza Epidemic', *Pall Mall Gazette* (9 December, 1889), 4.
24. 'This Morning's News', *The Leeds Mercury* (9 December, 1889), 5.
25. 'Austro-Hungary', *The Times* (10 December, 1889), 5.
26. 'Our London Letter', *Northern Echo* (11 December, 1889), 3; and 'The Influenza Epidemic at Home and Abroad', *Pall Mall Gazette* (11 December, 1889), 4.
27. 'London: Friday, December 13, 1889', *The Times* (13 December, 1889), 9.
28. 'Foreign and Colonial News', *The Times* (14 December, 1889), 5.
29. 'The Epidemic of Influenza', *The Times* (17 December, 1889), 5.
30. 'The Epidemic of Influenza', *The Times* (18 December, 1889), 5.
31. 'The Influenza Epidemic', 4.
32. 'To-Day's Tittle-Tattle', *Pall Mall Gazette* (1 January, 1890), 5.
33. Lori Loeb, 'Beating the Flu: Orthodox and Commercial Responses to Influenza in Britain 1889–1919', *Social History of Medicine*, 18.2 (2005), 203–24.
34. Loeb, 'Beating the Flu', 211–13.
35. 'Has Influenza: How to Cure it and How to Prevent it', *Pall Mall Gazette* (9 January 1890), 1.
36. 'THE COMING EPIDEMIC! THE COMING EPIDEMIC!!!', *Illustrated London News* 96 (1890), 31.
37. 'The users of SALT REGAL have hitherto escaped THE EPIDEMIC', *Illustrated London News* 96 (1890), 96.
38. F.B. Smith, 'The Russian Influenza in the United Kingdom, 1889–1894', *Social History of Medicine*, 8.1 (1995), 55.
39. David K. Patterson, *Pandemic Influenza, 1700–1900: A Study in Historical Epidemiology* (Totowa: Rowman and Littlefield, 1986), 1, 91.
40. F.A. Dixey, MD, *Epidemic Influenza* (Oxford: Clarendon Press, 1892).
41. P.J. Waller, *Town, City and Nation: England, 1850–1914* (Oxford: Oxford University Press, 1983), 25; and Smith, 'The Russian Influenza in the United Kingdom', 57.
42. 'The Influenza', *Spectator*, 66 (1891), 719; and Alfred Schofield, MD, 'A Few Notes on Influenza', *Leisure Hour*, 40 (1891), 570.

43. 'London: Saturday, December 21, 1889', *Lancet*, 134 (1889), 1293.
44. Patterson, *Pandemic Influenza*, 68.
45. Smith, 'The Russian Influenza in the United Kingdom', 61–65.
46. James Payn, 'Our Notebook', *Illustrated London News*, 96 (1890), 98.
47. Andrew Wilson, 'Science Jottings: Our Monthly Look-Around', *Illustrated London News*, 96 (1890), 146.
48. There was a comic song entitled 'The Influenza' published in 1890. Composed by George Dance and arranged by Fred Eplett, it recounted a number of the notable features of the pandemic (sneezing, people thinking they would escape, drinking) before listing the beneficaries of the influenza: the press, the doctors, the chemists, the drapers (on account of the demand for handkerchiefs) and, of course, the writers of comic songs.
49. Wilson, 'Science Jottings', 146.
50. Florence Fenwick Miller, 'The Ladies Column', *Illustrated London News*, 96 (1890), 154–55.
51. Miller, 'The Ladies Column', 154.
52. Miller, 'The Ladies Column', 154.
53. Miller, 'The Ladies Column', 154.
54. Morrell McKenzie, 'Influenza', *Fortnightly Review*, 55 (1891), 881.
55. McKenzie, 'Influenza', 882.
56. N. Katherine Hayles, 'The Materiality of Informatics', *Configurations*, 1.1 (1993), 154.
57. 'London, Tuesday, December 3, 1889', 9.
58. Peter Broks, *Media Science Before the Great War* (London: Macmillan, 1996), 7; Richards, *The Commodity Culture of Victorian England*; and Patrick Brantlinger, 'Mass Media and Culture in Late Nineteenth-Century Europe', in Mikulàš Teich and Roy Porter (eds), *Fin de Siècle and Its Legacy* (Cambridge: Cambridge University Press, 1990), 98–114.
59. 'Influence', *Great Thoughts*, 1 (1884), 34.
60. Gillian Beer, 'Wave Theory and the Rise of Literary Modernism', in *Open Fields: Science in Cultural Encounter* (Oxford: Clarendon Press, 1996), 295–318; Roger Luckhurst, *The Invention of Telepathy, 1879–1901* (Oxford: Oxford University Press, 2002); Pamela Thurschwell, *Literature, Technology and Magical Thinking, 1880–1920*, (Cambridge: Cambridge University Press, 2001); and Warner, *Phantasmagoria*, 253–84.
61. Thurschwell, *Literature, Technology and Magical Thinking*, 37–64.
62. Harold Bloom, *The Anxiety of Influence: A Theory of Poetry* (Oxford: Oxford University Press, 1973), 95. My thanks to Andrew Eastham who brought these remarks to my attention.
63. N. Katherine Hayles, *How We Became Posthuman: Virtual Bodies in Cybernetics, Literature, and Informatics* (Chicago: University of Chicago Press, 1999), 25–49.

9
Linguistic Trepanation: Brain Damage, Penetrative Seeing and a Revolution of the Word

Laura Salisbury

On 2 August 1915, the Russian-born American author John Cournos found himself lamenting the fragile materiality of the skull in an obituary for the Vorticist sculptor Gaudier-Brzeska:

> . . . and Brzeska is really dead, a vorticist with a vortex. He was vorticism. His fine mind had mastery over matter but in the end the foe won. A German bullet, that small but efficient vortex of materialism crashed into his skull – vortex interpenetrating vortex – and created a void which can only now be realized by those who knew him and enjoyed his friendship.[1]

Although Gaudier's 'fine mind' had mastery over the materials of sculpture, it remained, in the end, no more nor less than brittle untranscendent matter – shattered brain and bone. The 'efficient vortex of materialism' of the bullet that created a void in the vertex of the skull necessarily reshaped and reimagined that space as one of loss and mourning, a vortex of material limit and impossibility, rather than of creative agency.

Only a few years after Gaudier's death, however, the wounded Guillaume Apollinaire finds, or at least imagines, that he is not creatively disabled by his head wound; instead, he is inspired by the X-ray of his pierced and surgically trepanned skull to link poetic innovation with new experiences and visions of embodiment. He proclaims: 'But is there nothing new under the sun? What! My head has been X rayed. I have seen, while I live, my own cranium, and that would be nothing new?'.[2] In 1920, T.S. Eliot, rather more famously, seems to use a shadowy image of an X-ray to articulate the longstanding capacity of art to penetrate the surface of the body in 'Whispers of Immortality'.

179

The poem determines that 'Webster was much possessed by death / And saw the skull beneath the skin' (I.1–2), whilst Donne too 'knew the anguish of the marrow, / The ague of the skeleton' (I.13–14).[3] Here, the use of the poetic instrument to see beyond the flesh to bone speaks of a literary tradition that holds its own longevity as a parallax effect that reveals life's speedy dissolutions. Poetry sees into the body to speak as a *memento mori*. But for Apollinaire, seeing the living skull offers a new vision of human subjectivity alongside a new possibility for poetry. The penetration of the body – through wounding and by those radiant waves of the X-ray that produce an image of the skull beneath the skin – becomes a sign of modernity, of reconfigured poetic possibilities, and of life, rather than tradition and the inevitability of death.

So Apollinaire goes on to suggest that the human subject of modernity is exposed to new technologies – machines – that will alter both the content of poetry and the means of representation:

> The air is filled with strangely human birds. Machines, the daughters of man and having no mother, live a life from which passion and feeling are absent, and that would be nothing new?

> Wise men ceaselessly investigate new universes which are discovered at every crossroads of matter, and there is nothing new under the sun?[4]

The contiguity of the private life of anthropomorphized machines, scientific investigations, and the new visions of human embodiment in Apollinaire's account of the reinvented subject of poetry, enables trepanned corporeality to be placed firmly at one of these 'crossroads'. This chapter thus takes Apollinaire's suggestion for investigating 'new universes' at face value: it explores what it means to read the opened up cranium as a meeting place, a junction, that describes and enables a particular historical moment to consider how previously unseen configurations of spirit and matter, mind and machine-body, might articulate themselves to forge new linguistic possibilities.

Language matters

For Apollinaire, then, the trepanned skull becomes a site of interpenetration carved out by the force of war that brings minds, bodies and machines into a commerce with one another that initiates linguistic innovation. But this rendering of the boundaries between cognition, expression, corporeality and technology threateningly and thrillingly permeable also seems to repeat an earlier moment within the history of modernity, language and the human. Fifty-five years previously the human machine had seemingly finally been divested of the immaterial murmurings of a ghost inside. In 1861 Paul Broca delivered a paper to the Société d'Anthropolgie in Paris detailing the case of a patient called Leborgne, who would come to be known as 'Tan'; in so doing, he established the first fully materialist model of language production to be accepted as scientific orthodoxy. Tan was given his soubriquet because, mysteriously, and for twenty-one years, it was the only syllable he had been able to utter. When his brain was examined by Broca at autopsy, however, a determining cause for 'Tan' was found. Leborgne's brain was revealed to have a lesion in the third frontal convolution of the left hemisphere caused by a cyst – a disturbance in an area of the cortex that was consequently inferred by Broca to be 'the faculty of articulated language'.[5]

Joining clinical description with pathological anatomy, Broca, and those who immediately followed him, produced complex morphological studies of the structure of the brain and its 'centres' upon which clinical models of the spatial localization of 'functions' such as language were then built. With his work *The Symptom-Complex of Aphasia*, published in 1874, Carl Wernicke worked to determine the spatial relation between anatomical brain centres and function, mapping the production of linguistic ability by laying it directly over the anatomical location of material centres, and then exploring the ways in which these centres might communicate. Wernicke offered the most influential account of localized brain function in the period by using descriptions, supported by diagrammatic representations, that reformulated language, as an agent of consciousness, as a material process of mechanical communication between 'centres' of sensory-motor units along 'association fibres'. Determining that aphasia resulted from a disruption of the 'psychic reflex arcs' used in normal speech processes and a disturbance of the pathways between the sensory memory images and motor movement images that created associations, the brain damage traced out by Wernicke's classical doctrine of aphasia, and the insights into function it threw into relief, offered a way of conceptualizing what was persistently to be referred to in the period as the breakdown of a decidedly modern human machine. Wernicke indeed emphasized the mechanical qualities of brain function by comparing damaged cerebra to the malfunctioning telegraphs of modernity, noting that '[i]f certain letters are missing in the apparatus, specific errors would be consistently repeated in the message'.[6] But this account of brain centres mechanically communicating and rendered visible through static diagrams produced a theory that turned language into what Sigmund Freud dismissively termed 'a cerebral reflex'[7] – merely a mechanical-electrical discharge that seemed, somehow, denuded of its living, perhaps even its specifically linguistic, qualities.

As John Forrester explains, there was, in fact, only a '*covert* theory of linguistic function' at work in Wernicke's system, for it was 'modelled upon the reflex theory of the nervous system, which was itself subtly parasitic upon the doctrine of the association of ideas'.[8] In 1829, Thomas Carlyle had aligned the discursive structure of associationism with the 'Age of Machinery',[9] affirming, with heavy irony, that for associationists the Universe was 'one huge, dead, immeasurable Steam-engine', with humans themselves reduced to 'a mere Work-Machine, for whom the divine gift of Thought were no other than the terrestrial gift of Steam is to the Steam engine; a power whereby cotton might be spun, and money and money's worth realised'.[10] Late eighteenth and early nineteenth-century associatonism influentially determined that units of thought – mental atoms – were, in Anne Harrington's terms, simply '"copies," "pictures," or "representative images" of direct experience' which interacted with one another according to fixed mechanical rules such as those of 'contiguity' or 'similarity'.[11] By importing the 'mental atoms' of associationism into his account of 'sensory-motor memories' transmitted according to the electrified mechanics of the reflex arc, Wernicke thus produced a theory of thought and speech in which mind and brain spoke in consonance with one another according to the paradigm of *l'homme machine*. But if early aphasiology suggests a retrenchment of mechanistic models of mind and body, there is another narrative emergent at the end of the nineteenth century and continuing into the twentieth that remains obsessed with the idea of seeing into the head to map the material production of thought and language, but which rewrites the ways in which the relationships between mind, body and machine might be imagined.

Seeing in

On 2 October 1910, Thomas Edison told Edward Marshall from the *New York Times* that even in the face of William James's alleged postmortem reappearance – a spiritualist event with which the newspapers had apparently been 'teeming' – he could not be persuaded to believe in the immortality or even the existence of the soul. Instead, Edison presents the notion of mind, which produces and underpins the idea of soul, as a mere epiphenomenon of a material, finite, brain: 'The brain immortal? No: the brain is a piece of meat-mechanism – nothing more than a wonderful meat-mechanism'.[12] Suggesting a penetration of neurological discourses into wider scientific and popular culture, Edison's proof of the mechanization of both mind and brain (which are elided here) is found, once again, in aphasiology's discoveries, whilst the simile used to explain brain function intriguingly relies on Edison's own invention – the modern machine of the phonograph:

> This brain of ours . . . is a queer and wonderful machine. What is known as the fold of Brocca [sic], at its base, is where lie stored our impressions in the order in which they are received. There, for instance, is where our knowledge of our mother tongue is stored . . . just as if that part of the brain were the particular phonographic cylinder on which it had been recorded.[13]

He then goes on to cite the case of a man injured in the 'lower part of the fold of Brocca [sic]', who had lost his mother tongue of English, but retained his later acquired knowledge of French and Greek. Here, in this profoundly topographical and spatialized model of brain and mind, language, thought and memory – those bastions of the subject's sense of its own transcendent inwardness and the stuff on which ideas of the 'soul' might feed – are all represented as fascinating, complex, but in the end merely material functions of a modern 'meat-mechanism'.

Edison's easy assimilation and reproduction of a mechanized, spatialized, model of brain function, visualized in phonographic terms, was, of course, part of more general discursive environment in which bodies could be imaged as fleshly machines. Hermann von Helmholtz had discovered in 1847 the first law of thermodynamics which determined that 'the *quantity of force which can be brought into action in the whole of Nature is unchangeable*, and can be neither increased nor diminished'.[14] In other words, *all* forms of energy – whether within the living or the inorganic – could potentially be transformed into one another. But if the forces within organisms could be explained using the same principles as those which applied to inanimate objects, there was no need to appeal to anything transcendental in order to explain the workings of 'life'. As Harrington reveals, in 1858 Rudolf Virchow asserted that bodies, alongside everything else, could be understood and conceptually transformed into the empirical measurability of the industrialized machines of modernity:[15] '[T]he same kind of electrical process takes place in the nerve as in the telegraph line [...] the living body generates warmth through combustion just as warmth is generated in the oven; starch is transformed into sugar in the plant and animal just as it is in a factory'.[16] Clearly, then, Edison is simply repeating the discourse of late nineteenth-century positivism in speaking of the brain as a 'meat-mechanism' powered by electrical impulses.

But in visualizing the brain as a repository, an archive, of phonographic cylinders – material objects laid out before the eye – Edison is also repeating aphasiology's own obsession with seeing beyond the occlusions of the cranium and making records of what it found there. As all histories of neurology make clear, the first theories of brain localization were profoundly indebted to phrenology's visible mapping of inferred organs of capability and capacity on to the bumps and depressions of the cranium.[17] It was Broca, however, who finally allied phrenology's insistence on the material location of subjective qualities with clinical observation and the study of morbid anatomy. It was Broca who opened up the skull post-mortem in order to see and to map losses of faculties on to lesions observable within the anatomical structure of the brain. He thus created the discipline of aphasiology as something that took part in and formed an extension of what Michel Foucault calls, in *The Birth of the Clinic*, the nineteenth-century 'technique of the corpse'.[18]

Foucault describes the emergence of a positivist modern clinical medicine in the late eighteenth century that overwrote the discourse of passing, transversing, sympathies, correspondences or homologies, by which disease previously fluttered or tore through the body. With the birth of the clinic there was a new concentration on observing the body's discrete functions and on extending the reach of observation by penetrating the skin and uncovering to the medical gaze solid organs – shapes of corporeality and of pathology – that were previously below the threshold of the visible. Significantly, Foucault picks as a fundamental exemplum of this new method a positivist brain pathology subtended by a penetrative mode of observation. Foucault describes the work of Bichat, the founder of modern pathology and histology, who carefully broke the cranium at autopsy in order to see the pristine brain inside, thus carving the way for Broca's discoveries later in the century:

The fruit is then opened up. From under the meticulously parted shell, a soft, greyish mass appears, wrapped in viscous, veined skins: a delicate dingy-looking pulp within which – freed at last and exposed at last to the light of day – shines the seat of knowledge. The artisanal skill of the brain-breaker . . . the precise, but immeasurable gesture that opens up the plenitude of concrete things, combined with the delicate network of their properties to the gaze, has produced a more scientific objectivity for us than the instrumental arbitrations of quantity. Medical rationality lunges into the marvelous density of perception, offering the grain of things as the first face of truth, with their colours, their spots, their hardness, their adherence.[19]

Importantly, Foucault suggests that this was not a gaze dependent on ideas of light and transparency that putatively revealed the essence of objects illuminated according to their ideality; instead, he argues, at the end of the eighteenth century and during the emergence of the modern anatomico-clinical method, to see was precisely to be faced with the recalcitrance of matter – a screen that obscured in order reveal its own shape and material surface. To see the surface of an organ, or to reveal the internal surface of a tissue sample, meant precisely not to see through it, but to trace over it with the eye in a way that illuminated and also produced the solid contours and textures of an object.

As Foucault explains, 'the solidity, the obscurity, the density of things closed in upon themselves, have powers of truth that they owe not to light, but to the slowness of the gaze that passes over them, around them, and gradually into them, bringing them nothing more than its own light'.[20] Clinical experience and indeed the clinical gaze produced samples, organs, and even individual bodies, which assumed a strangely calcified solidity, then: 'No light could now dissolve them in ideal truths; but the gaze directed upon them would, in turn, awaken them and make them stand out against a background of objectivity'.[21] To penetrate the body and to see in this way was thus to begin the endless labour of empirical observation and description. If, however, to see meant penetrating the skin or depending on the 'artisanal skill' of the bone breaker, then the most stable, the most clearly observable manifestations of disease localized in particular organs and bodies, necessarily appeared post-mortem. It was death that allowed the clinician 'to separate functional complexity into anatomical simplicity'.[22] Denuded of life, of the complex movements, combinations and interpenetrations of function, the corpse rendered organs visible in their singularity, mirroring the way in which disease itself interfered with the functional flow of the living body as organism, separating off a damaged organ from its purpose as part of a wider functional system. As Foucault puts it: 'Anatomy could become pathological only insofar as the pathological spontaneously anatomizes. Disease is an autopsy in the darkness of the body, dissection alive'.[23]

It is significant, however, that the language of the clinic is smoothly reflective of this penetrative gaze, for Foucault. Language indeed becomes a way of rationally verbalizing and then communicating the contours of stable, observable organs, and the solidity of experience, as saying is elided with seeing. Paradoxically, the 'loquacious gaze' of the positivist clinician uses an idea of language which seems, at least for twentieth- and twenty-first-century readers, so clear, and so consonant with what is seen, that it seems to hide behind a mask of transparency its textuality, its imbrications with internal self-difference and those representational displacements with which words necessarily seem always to become contaminated. If there is a tendency for positivist medicine in this period to use a language that imagines it slides so smoothly over the visual that it seems to reflect 'the non-verbal conditions on the basis of which it can speak',[24] it becomes revealing and ironic that in Wernicke's classic doctrine of aphasia, the character of language as a living function is fundamentally overwritten by the stolid visibility of the brain as a static organ in an opened-up corpse. As the anatomico-clinical method translates 'functional complexity into anatomical simplicities',[25] language is denuded of what might be thought of as its 'living' qualities. At this historical moment, then, language in aphasiology seems to assume two contradictory positions. It coagulates as the secretion of solid brain matter; but the materiality of language also dissolves in the translucency of an empirical description in which it becomes little more than a lens for rendering that matter visible.

Dissection alive

So it is revealingly appropriate that the famous nineteenth-century neuroanatomist and psychiatrist, Paul Flechsig, should have chosen to have his 1909 Festschrift photograph – the culminating image of his career – taken surrounded by slices and samples of cerebral matter and in front of a picture of an all-dominating and partially dissected brain (see Figure 10). By the 1890s, however, the discipline of aphasiology had already begun to press against the limits of this 'technique of the corpse' as it searched for new passages into the living brain. Most neurologists exploring theories of localization had, of course, relied upon the liveness of animals – vivisection – to demonstrate and extend their postulations. Using an instrumental mode through which the animal became conceptualized as a usefully animated corpse, the neurologist produced brain lesions that spectacularly projected on to the visibility of animal corporeality a map of the brain's localized motor and sensory functions.

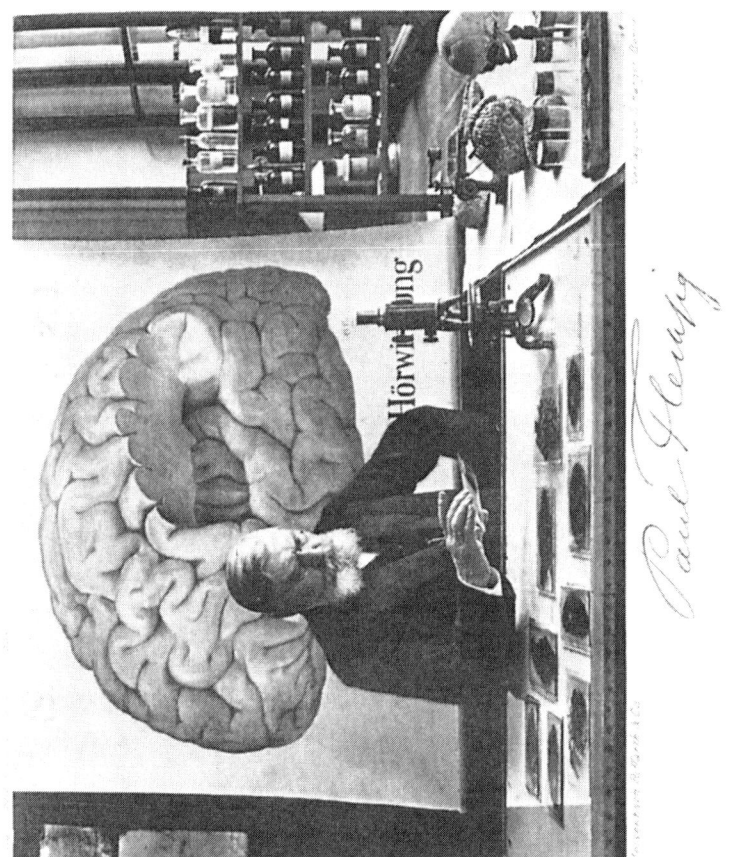

Figure 10: Paul Flechsig (1909)

But if the human has been singularly defined as the animal that talks, as in Aristotle's classical definition of *zoon logon echon*, there was no vivisection that could be ethically performed that would reveal much about diminished linguistic capacity. As we have seen, faced with the silence of the corpse and the wordlessness of animals, neurologists like Broca and Wernicke relied solely on the natural experiments of disease or wounding to produce observable symptoms in living humans, which could only later be matched up with morbid anatomy.

But the emergence of Wilhelm Röntgen's X-ray imaging technique in 1895 seemed to offer a new technology which could penetrate the body painlessly, and open it up to the gaze without the need for it to be reduced to a corpse. The X-ray offered to turn the body inside out without surgical penetration, to image and imagine interiority on what were now visible external surfaces seemingly without damaging that which it sought to observe.[26] So it is no surprise that less than three months after Röntgen's discovery, William Randolph Hearst was telegraphing Edison with a message. On 5 February 1896 he wired: 'WILL YOU AS AN ESPECIAL FAVOR TO THE JOURNAL [*The New York Journal*] UNDERTAKE TO MAKE A CATHODOGRAPH OF HUMAN BRAIN'.[27] Edison took up the challenge, boasting that he would soon be able to transcend the recalcitrance of the cranium, but he acknowledged defeat only a few weeks later. X-rays, it turned out, were not very useful for offering up images of soft tissue, as solid organs 'dissolved' alongside the flesh in a strange literalization of Foucault's notion of the transparent gaze of ideality.[28] Nevertheless, the excited fantasies that the medical gaze would soon be able to see the living, working brain without damaging the protective casing of the skull, insisted both in medical and in popular cultural discourses. In 1905, William Shuler Harris's science fictional musings on *Life in a Thousand Worlds* imagined that one of the features of the ultra-modernity of the planet Ploid was its development of 'thought photography'. He details

> their ability to follow the course of thought in a living cerebrum after the brain has been made visible by a light more potent than the X ray. After this exposure the operator, with his wizard magnifying lens, watches the tiny tremulous brain cells in their infinitesimal quivering, as they carry messages from the soul to the world of sense and being.[29]

For Harris, the limits of visibility of brain function are only drawn by the capacities of technology. Of course, it was not until the 1970s that the emergence of positron emission tomography (PET), and then functional magnetic resonance imaging in the early 1990s, enabled neurobiological correlates of human *behaviour* (rather than brain structure) to be mechanically reproduced, even though it remains uncertain still precisely how those changes of blood flow in the brain registered in brain imaging techniques are orchestrated to serve brain activity.[30] Thought photography or endopsychic perception was and indeed remains a summit against which brain imaging is triangulated, rather than something that has ever been achieved.

So in the face of the technological limitations of the 1890s, the problem of how to map and analyse brain function still found its clearest articulation in the displaced mappings of aphasia. And it was the neurologist and nascent psychoanalyst Sigmund Freud, rather than Röntgen or Edison, who in 1891 began to suggest the possibility of another way of getting into the head. In *On Aphasia*, Freud opposed himself to the hegemony of 'localization' by complicating the tendency, since the work of Ludwig Lichtheim in 1885, to hold as distinct aphasic symptoms caused by lesions of the language 'centres', and those attributable to the destruction of neural pathways. He describes how in most aphasiological writings, there is an assumption that 'the speech apparatus consists of distinct centres separated by functionless areas, and that ideas (memories) serving speech are stored in certain parts of the cortex called centres while their association is provided exclusively by subcortical fibre tracts'.[31] Freud, however, found no anatomical evidence for this. In opposition to this schematic but visibly explicable idea of a nineteenth-century machine, Freud determines that '[n]o single individual nerve fibre and nerve cell is enlisted for a single language association function'; rather, 'the speech area is a continuous cortical region within which the associations and transmission of underlying speech functions are taking place'.[32] He indeed insists that such dynamisms of function 'are of a complexity beyond comprehension'[33] – beyond being conceptualized as a telegraph, typewriter or phonograph, and certainly beyond any static modelling in a diagram.

In Freud's new model, the cortex becomes both more homogeneous and more functionally malleable. No mere mechanical repository of discrete but connected material spaces as one might expect of a machine, the brain disperses its functions across the cortex, as matter adapts its function to changing conditions. Freud notes that because of the plasticity of brain matter and function, lesions can affect processes in areas of the cortex far removed from the wound; similarly, and by implication, new modes of association, new productions of meaning, can reform in a brain that has been damaged. Perhaps the most radical innovation of Freud's model is the fact that it explicitly brings the vagaries of representation into the materiality of brain function. Freud uses the example of the 'body-schema' (the non-conscious system through which the body understands itself to be spatially organized, or how one part of the body relates to another) to illuminate the fact that all perception processed in the brain is combinatory in character – a moulded, displaced 'representation' rather than the simple mirroring or reflective 'projection' in which 'the periphery of the body is contained in the cerebral cortex . . . point by point'.[34] Against the prevailing doctrine of simple associationism, he writes that the fibres connecting the periphery of the body with the cortex do not just convey, say, 'retinal stimulus'; instead, they combine a number of different impressions – in this case retinal impressions with kinaesthetic ones – in order to produce meaning:

> We can only presume that the fibre tracts, which reach the cerebral cortex after their passage through other grey masses, have maintained some relationship to the periphery of the body, but no longer reflect a topographically exact image of it. They contain the body periphery in the same way as – to borrow an example from the subject with which we are concerned here – a poem contains an alphabet, i.e. in a completely different arrangement serving other purposes, in manifold associations of the individual elements, whereby some may be represented several times, others not at all.[35]

What Freud asserts here is that perceptual data do not stand alone as mirroring projections of the 'mental atoms' associated with perception; they are, instead, combined in representations which function like a language.

Just as meaning in language is not held solely or even mostly within individual words, but within structural relationships between signs and more broadly syntactical relationships, it is the 'manifold associations' that hold the valuable neural information. So, as Forrester puts it, here '*all* representations are coded *as if* they were a language'.[36] In terms of Freud's structural analysis of the word itself, its material, neurological nature is also revealed to be a startlingly sophisticated combination or 'complex of concepts (impressions, images) which through its sensory part (its auditory component) is connected with the complex of object associations'.[37] For Freud, then, all aphasias are reconfigured as disorders of conduction, combination, or association[38] – disruptions of the structural relationships between 'word-presentation' and 'object associations' – rather than disturbances of centres, or of the nerve pathways between centres. Freud indeed affirms that '[m]ost lesions are not directly destructive and they have a disturbing effect on a much larger number of nervous units than those immediately involved'.[39] Consequently, disruptions are occasioned by a cerebral lesion, but it is not the lesion that 'speaks' in the symptom; it is the complex disorder of association, the disturbance and remaking of the ways in which perceptions are combined in representations to make meaning, that the doctor must 'hear'.

Where saying was merely a smooth reflection of what can be seen in the 'technique of the corpse', here all representation – whether directly linguistic, or bodily-schematic – needs to be rethought according to the idea that meaning is made within and between combinations of associated elements, and is dependent on the general structural functioning of an overarching system. So for Freud, the functions, modes of association and shifts in the structural processing made visible in the aphasic symptom all now have to be interpreted according to a strong theory of how meaning is constructed in the brain as a complex process, and how it can then be reconstructed to form a legible symptom for the doctor. Freud's psycholinguistic investigation of aphasia, which determines the importance of tracing out the shapes of impaired language in ways that take account of its own principles of organization and combination, thus finds textual description rather than a diagram to be aphasia's most accurate representational tool. As L.S. Jacyna puts it: 'While for classical aphasiology the disease was to be understood as the perturbation of a timeless extended mechanism best represented by a diagram, now it was seen as explicable only in terms of the epic narrative of the unfolding nervous system'.[40]

So language is not simply to be found in the action of one particular area of the brain-machine as it had been presented in classic localization theories; language is instead produced according to processes of brain activity that can themselves only be adequately represented and explained through a temporal narrative in which function unfolds.[41] One version of the emergence of psychoanalysis indeed imagines a story of Freud declining his position as a neurologist, if a neurologist is a diagram-maker of brains analysed at autopsy. He rejects the diagrammatic fervour of the 1880s in favour of diagnostic listening and writing that refuses the primacy of what can be empirically seen, but tries to hear and interpret shapes of arrangement emergent from an orifice in the head that is found in life rather than made after death – a mouth uttering language.[42] Psychoanalysis as a practice has indeed been framed as a turn from visual modes of representing and categorizing mental illness as a form of organic, neurological pathology, towards a method that valued language and the structures of representation above any empiricist biologism. Sander Gilman writes: 'Freud rejected the idea of seeing the patient, thus centring psychoanalysis on the process of listening'; and 'in rejecting the rigid representationalism of nineteenth-century theories of understanding mental processes, Freud also rejected their basis of empirical proof'.[43]

Importantly, though, the practice of psychoanalysis has always been more than simply diagnostic; in Anna O's famous suggestion to Breuer, it emerges as the 'talking *cure*' for hysterical symptoms.[44] It is, at its core, a temporal process that consists of re-evoking a memory of trauma and subjecting it to a reverbalization that reconnects the event with its disarticulated affect. In a revealing reversal of aphasic brain damage, psychoanalysis is a process of finding a way to articulate that reforms the pathways between 'object associations' and 'word presentations' which have become dissociated in psychological trauma and written on the body as a speechless symptom. In a letter to Wilhelm Fliess in 1896, Freud himself suggested that it was in his aphasia book that he offered the first description of nerve pathways subject to 'rearrangement' or 'retranscription', and that this formed the core understanding of both how the psyche works, and how psychoanalysis might effect its cure: 'the material present in the form of memory traces [is] subjected from time to time to a *rearrangement* in accordance with fresh circumstances – to a *retranscription*,' he writes.[45] It is this process that allows the repression of the cause of the trauma following a psychological wound – the creation of what Forrester calls 'the lesion of an idea';[46] but it is also what allows the analyst to re-present the initial trauma, and for the relationship between memory and affect to be reforged using words. Psychological disconnections and reconnections thus follow the same structural forms as those found in material brain damage and brain healing, as the relationship between particular 'word presentations' and 'object associations' can be disturbed within a damaged brain and functionally reconnected in relation to experience. The 'linguistic' method of gaining a diagnostic and curative access to the structure of brain and psyche, of course, emerges in the face of the material impossibility of 'thought photography' or a neurosurgery that could suture the connections lost in aphasic disturbance. For although successful brain surgery became increasingly common in the late nineteenth century because of the belief in the value of antisepsis and a basic, though vital, sense of which localized areas of the brain must be avoided in operations,[47] neurologists knew enough about the complexity of the localization of the language that few attempted to relieve aphasia using the surgeon's knife and suture unless a tumour or abscess was immediately life-threatening.[48] Consequently, at this historical moment, words were not just the best tools for describing and imagining the shapes of neurological and psychological disturbances of association, they became the only instruments that could safely 'open up' the cranium to expose and then knit together the brain's disrupted connections.

Faced with the mind of the hysteric or neurotic, or even, one might infer, the alive brain of the aphasic patient, it is the word, then, that creates a passage into the recalcitrant matter of the cranium; it is through words that the interiority of mind and brain is revealed and, potentially, reconfigured. In 1890, Freud indeed affirmed in an article for a popular medical textbook the importance of words as both diagnostic tools and as reparative instruments – seemingly immaterial forms that could 'magically' reshape the functioning of brain and mind beyond the topography of nineteenth-century medical visibility:

> Words are the essential tool of mental treatment. A layman will no doubt find it hard to understand how pathological disorders of the body and mind can be eliminated by 'mere' words. He will feel that he is being asked to believe in magic. And he will not be so very wrong, for the words which we use in our everyday speech are nothing other than watered-down magic.[49]

Words become diagnostic instruments of penetration and revelation, but they also 'eliminate' pathology, enabling restoration at a historical point where the surgeon's instruments and imaging techniques meet their material, although not metaphorical, limits in relation to living bodies and minds.

So what are the shapes of language and brain function that late nineteenth-century aphasiology reveals? As we have seen, Foucault describes how disease performs live dissections on the body. Disease displays organs that are darkly solid – discrete and impenetrable – as corporeality is denuded of the sense that it is formed of interpenetrating spaces through which processes flow. It seems, then, that aphasic language mirrors this diseased body, in its obdurate and frustrating solidity. For as with Leborgne's 'tan' speech automatism, aphasic language refuses the associative suppleness through which it could become so translucently at one with mind that its material character would seem to dissolve. Pathological language is clinically singular, however, because it becomes a solid, lifeless lump of matter – it presents itself as a diseased object – at precisely the moment when medicine also needs to assert a language that is transparently at one with the gaze of the clinician as subject. And this resistance produces sufficient friction that the transparent medium is forced into self-reflection. In this moment of disjunction, where language as a pellucid medium and language as material object meet, words are deformed to assume a reflexive shape, becoming object, subject, and the passage between them. Language is no longer a translucent medium in which seeing and saying become one in the occulting transparency of those modes of medical representation aligned with positivism. At the beginning of the twentieth century, language, as it emerges from neurology and psychoanalysis, becomes instead a complex surface that is so scored and scratched that one can no longer see through it. Of course, its surface is scuffed up precisely because, as subject and object, it is continually in disjunctive, resistant contact with itself; and it is within these reflexively abraded lesions that it begins to 'speak'. The quality, the living texture, of language indeed begins to extrude as it reveals itself as always already implicated – never simply object, subject or a transparent copula – but a complex surface that must be traced through time rather than mapped diagrammatically or projected as a static image. And perhaps, then, it is precisely from the 'crossroads of matter' emergent in the aphasic symptom that language assumes the shape of a complex, roughened-up surface that reveals and reflects consciously upon itself. Aphasiology indeed traces out a specific context within which the particularly material sense of language as a diagnostic tool, a reparative instrument, and the expression of a wound, which insistently haunts hopes for a 'twentieth-century word', seems meaningfully to speak.

Revolutionary words

There is a banner headline: 'GIRL, OPERATED ON FOR AMNESIA, SPEAKS 12 LANGUAGES'. Originally a story from the Paris edition of the *New York Herald*, the article goes on to detail a slightly different set of circumstances:

> Mlle. P.M. . . . was suddenly struck with amnesia following the puncture of her pleural cavity. A radioscopic examination revealed nothing pathologically wrong beyond traces of a pulmonary tuberculosis. After a deep sleep she woke 'like a person transported to an unknown country'.

Originally remarkable for her intelligence, she now could not even express herself in her native French . . . Her hands could not use the simplest objects. Finally, several weeks of instruction taught her to count, but she still could not understand the simplest explanations.

Then, suddenly, 'the gift of tongues' was bestowed upon her. She began speaking twelve different languages, none of which she had known before her amnesia. She wrote them too, with her left hand, which she had never used before. But why?[50]

The most obvious medical answer to the question is that the procedure of thoracentesis – the puncturing of the thorax with a fine needle to obtain a sample of liquid from the pleural cavity for diagnostic purposes or the alleviation of pressure – can produce, as one of its side affects, acute cerebrovascular accident. A stroke would explain the motor dysfunction, the amnesia, and the aphasic symptoms of Mlle P.M.; and although there are no recorded cases of aphasic patients beginning to speak languages to which they have never been exposed, it is not altogether uncommon for aphasic patients to lose one of their languages and not another, or to have their speech fluency disturbed to the degree that it sounds as though they are speaking in a different accent. It is perhaps not so hard to imagine that journalistic licence could have shaped this neurological symptom into 'the gift of tongues'.

What is significant here is not that this case should have been reported and somewhat distorted; what is significant is that the story appears in a section called the 'Mantic News' of the periodical and modernist 'little magazine' *transition*, published in 1932. Nestled between accounts of 'medieval' curses still operating near Oldham, a magic potato that brings luck to a Texan Secretary of Labor, witchcraft in the Black Country, 'strange cases of nervousness' in Havana blamed on the actions of witch doctors, and the adoption of Rudolf Steiner's anthroposophic farming methods in England, the story makes the 'Mantic News' because it asks: 'When a surgical needle punctured her pleural cavity what strange reservoirs of the past did it draw upon?'.[51] Here, the penetration of the body and the ensuing symptom of brain damage is read as a conduit back to some form of primal knowledge, whilst, as part of the Mantic News, it presumably also traces out the shape of a future speaking subject. In the Mantic News, the aphasic patient, although subject to a wounding, also speaks a language that functions as a conduit to repair. Her language emerges as if from a hole burrowed into the past and one bored towards the shape of things to come.

The Parisian periodical *transition* was edited by the multi-lingual American journalist and writer Eugene Jolas and the Mantic News was almost certainly assembled by him. First published in 1927 and running in its initial form until 1938, *transition* committed to publishing Dadaist, surrealist and expressionist texts, alongside any work Jolas considered to be held in the pull of what Dougald McMillan has referred to as a 'wave of neo-romantic, irrationalist thought'.[52] *[T]ransition* is best known as a magazine that published major works by Gertrude Stein, and the place where James Joyce first made available sections of *Work in Progress* (later to be *Finnegans Wake*). Jolas, however, was more mystically minded than either Stein or Joyce, and his doctrine of the 'Revolution of the Word', which was published alongside and as part of a defence of *Work in Progress* in 1929, appeared persistently in *transition* as a mode which sought to remake the human subject through a formally innovative, revivified language which would act as a passage for the individual's intuitive, so-called subconscious, impulses.

In *transition* 15, Jolas outlined the crisis in both economic and spiritual values precipitated by the 1929 Crash and proclaimed the need for a new language that could both express and repair the current situation: 'The reality of the *universal word* is being still neglected. We need the twentieth-century word. We need the word of movement, the word expressive of the great new forces around us . . . The new vocabulary and the new syntax must help destroy the ideology of a rotting civilization'.[53] In the 'Proclamation of the Revolution of the Word', Jolas accordingly asserted the right of the literary creator 'TO DISINTEGRATE THE PRIMAL MATTER OF WORDS IMPOSED ON HIM BY TEXT-BOOKS AND DICTIONARIES', appealing persistently to the language of Romantic creativity and agitating for a 'pure poetry' that could represent a 'LYRICAL ABSOLUTE THAT SEEKS AN A PRIORI REALITY WITHIN OURSELVES ALONE'.[54] To effect these disintegrations of language, Jolas recommended the use of a roughened-up, defamiliarized lexicon – neologisms, portmanteau words, slang – and he persistently demanded the disruption of transparent syntactical laws. At each moment, he remained true to the notion that a poetic language bound to the expression of what was primal in the personality could liberate both the word and the human subject from the utilitarian restrictions of a society dominated by what he called 'fordian standardisation'.[55] He also kept faith, however, with the idea that the language of modernity had new experiences to which it needed to bear witness. So for Jolas, a reborn, reshaped language became both the effect of new forces and experiences and the tool through which innovation and a subjective, individual, revolution of word and soul might be both recorded and extended. In the face of gathering forces of Left and Right in the early 1930s, Jolas claimed that it was 'the poet in giving back to language its pre-logical function [who] makes a spiritual revolution – the only revolution worth making today'.[56] For him, only a remade language could express, with smooth symmetry, the interiority of what was universally human – 'the illumination of a collective reality';[57] and it was thus only a revolution of the word that could evoke a true internationalism of the spirit, a glottological unity, and a 'universal idiom' that could stand up to national and nationalist linguistic forms, 'bridg[ing] the continents and neutraliz[ing] the curse of Babel'.[58] On one level, then, Jolas seeks a language that might be defined as modernity's wound. For the tri-lingual Jolas, caught between national languages, it was indeed clear that, for him, '[l]anguage became a neurosis'[59] – a record of the loss of any primal, unitary language after Babel – just as in Mlle P.M.'s symptoms multilingualism was an indication of a lesion. But, significantly, the Mantic News tells us that Mlle P.M.'s wounded language, which extended its contours by seemingly absorbing other tongues and new usages, was also a conduit through which something universal might be restored.

This sense of language as both wound and weapon should, I think, be read as part of a determinedly postwar crisis in language and the formations of meaning. Jolas speaks of his return to Europe from the USA after the War as a 'release from technological pragmatism and machinism', and there is a strange sense in which the industrialized destruction in Europe seems, for him, to function as a kind of fortunate fall.[60] Vincent Sherry has argued that '[d]ead bodies in their millions weighed out the disproof of the nineteenth-century myth of progress through technology, that grand syllogism of liberal history', and Sherry identifies literary modernism's ensuing loss of faith in the restrained, colourless, language of liberal rationalism which seemed to march noiselessly alongside it.[61] For Jolas, certainly, there was the desire for a new word of extremity and irrationality – an extruding word – as the War screamed the need to cut through the smoothly obscuring 'hypnosis of positivism' and of the ideology of the machine, in all its multiple configurations.[62] Jolas indeed affirms in 1932 that 'creative expression today' is a 'cowardly genuflection before positivist dogmas and social idols[.] Progress and evolution as watch-words. Machine-mysticism as progress'.[63] Jolas thus demands a 'Revolution of the Word' that recognizes the 'primordial background of life . . . characterized by enormous scissions',[64] and can invoke a language bent on boring down, on 'encouraging the revolt of the sunken "I"'.[65]

It is significant that the surrealist André Breton, who published in *transition* in the 1920s, and Jolas were both writers who had been forced to face, first-hand, the capacity of modern, mechanized warfare to mass-produce new modes of articulation – new configurations of language that erupted from psychological trauma and the bullet's 'efficient vortex of materialism'.[66] Breton's surrealism and his *écriture automatique* was at least partially born from the noise of the psychotic and hysteric which spilled out of the neurological wards in which he worked during the War. Jolas also witnessed the material breakdown of language as a record of a wounded, reshaped subjectivity during his time as a secretary in

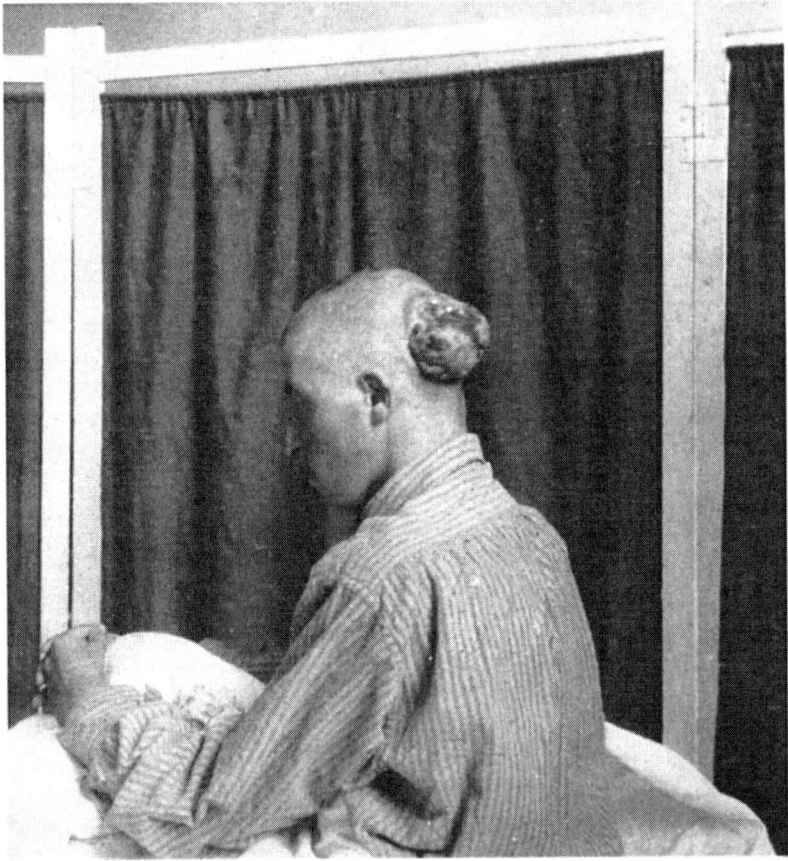

Figure 11: Military Hospital V.R. 76, Ris-Orangis, France: Soldier with head wound from battle at Verdun in First World War (1916)

a war-time neuropathic ward when he transcribed and typed up case-histories of wounded soldiers. In his own words, he 'had to listen to fantasies and hallucinations, to incessant weeping. I heard paranoiacs crying out against a hostile world',[67] and this experience steered Jolas towards the writing of poetry forged from a new language of unmediated contact with the irrationality of the exposed 'night mind' and those illegitimately opened-up crania of wounded soldiers (see Figure 11). In 'Notes on Reality' from *transition* 18, Jolas indeed states that the primitive and the irrational towards which the Revolution of the Word hopes to move remains uniquely accessible in and as particular forms of estranged, displaced, un-normative language which render suddenly and startlingly visible the opaque materiality of the word found beyond – and often violently beyond – the discourse of rational consciousness. For Jolas, 'the primitive mythos is a subterranean stream (held up by "civilized" consciousness) which we observe again and again in such manifestations as the dream, neuropathic conditions, and in the poetic inspiration as such'.[68]

Penetration is vertical

It is important to note that *transition* places itself precisely at what Apollinaire calls a new 'crossroads of matter' – a new sense of the relationship between mind, body and language – by explicitly opposing itself to associationism and late nineteenth-century theories of brain localization. Stuart Gilbert writes revealingly in *transition* 21's 'Art and Intuition' of the spiritual decline of modernity that is the devastating result of a reduction of thought to the firings of a mechanism. He deplores the vestiges of associationism that remain in discussions of consciousness that elide mind with the materiality of the brain and he articulates the hope for a theory of the intuitive immediacy of experience through which the subject can be aligned with the forces of the external world. 'We hear much talk of subjective and objective, as if consciousness were situated (like the brain) at a point of space or in an Eiffel Tower, and there received messages from an "outside world" and transmitted replies, "reacted",' he writes;[69] but he insists that the mind is not a machine which simply records passively data from the outside world but something which can achieve intuitive immediacy and unity with what lies outside itself. '*Give us back our dreams*,' he cries, 'despite the menace of machine-men and their machine-guns.'[70]

In an extraordinary article from 1932 entitled 'The Structure of the Personality' by the German Expressionist writer, doctor and one-time

Nazi sympathizer Gottfried Benn, an account of material brain function also emerges that becomes highly influential upon the 'Revolution of the Word'. Suggestively, Benn pits the discourses of the 'younger sciences' – the 'Theory of Types', the 'Theory of Expression', 'Gestalt Theory'[71] and 'Psycho-analysis' – which saw themselves as having a material basis even as they gestured beyond any biological reductionism, against a theory of human personality which has its roots in the localization-ist neurology of Broca and Wernicke.[72] Benn describes the influential orthodoxy of these localizationist theories as a 'kind of cranial theory [a phrenology] turned inward', and although he does not deny the fact that 'the hemispheres of the big brain . . . are doubtless the seat and center of action of the human intelligence', he writes dismissively of the idea of 'autonomous psychic forces', 'associations which passed through anatomical conduits', and the 'constructive, mathematical psychology' that emerged from 'the age of Flechsig and Wernicke'.[73]

Using scientific discourses borrowed from the neurologist John Hughlings Jackson, Benn suggests a form of brain construction in which geological stratifications of function are laid down over time, each of which are contained within and indeed accessible to the modern human. Following Benn closely, Jolas goes on to describe his own mystical yet persistently material model of brain construction and function in his 1932 monograph *The Language of the Night*. There, that which is most 'primal' in terms of brain function peeps tantalizingly from under the capacity for rationality, as though waiting to be released:

> The capacity of 'mystic participation' which, according to Lévy-Bruhl, exists in primitive mentality, survives in civilized man, albeit in atrophied form. The pineal body – a fragment of the exclusive cerebral organ of our ancestors – is still part of the brain structure. The growth of the big brain has smothered this unit which the ancients called 'the seat of the soul', 'le troisième oeil' [third eye], 'das Scheitelauge' [parietal eye].[74]

While Jolas speaks of a material covering of cerebral matter that must be penetrated, Benn argues similarly that modern man's 'big brain', which specializes in conscious, rational activity and mentation, sits on top of an earlier substratum, although it can be opened and passed through in the dream, drunkenness or psychosis. Through a new orifice bored into the cranium – not the mouth, but another hole in the head – something else can be seen and emerges to speak. In Benn's terms:

We carry the ancient peoples in our souls [. . .] When the logical super-structure is loosened, when the scalp, tired of the onslaught of the prelunar states, opens the frontier of consciousness about which there is always a struggle, then there appears the old, the unconscious, in the magical transmutation and identification of the 'I', in the early experience of the everywhere and the eternal. The hereditary patrimony of the middle brain lies still deeper and is eager for expression: if the covering is destroyed, in the psychosis there emerges, driven upward by the primal instincts, from out of the primitive-schiszoid sub-structure, the gigantic archaic instinctive 'I', unfolding itself limitlessly through the tattered psychological subject.[75]

Here, then, various forms of 'brain damage' are read, in accordance with Hughlings Jackson's theories of aphasic disturbance that Benn cites, as somewhat continuous with dissolution – a form of evolution in reverse that releases 'earlier' strata of brain function into the present.[76]

But in *transition* as a whole, it seems that it is a particular kind of artist, rather than simply the subject of a head wound (in all its multiple forms) or the clinician, who can and must use language to access these primal parts of the brain. Jolas begins to insist that language should not simply record the states of the 'tattered psychological subject' – the subject damaged by trauma, wounding or disease; instead, poetic language, functioning through a 'voluntary, mediumistic conjuration',[77] should work as a probe to bore down vertically through the big brain to release the 'gigantic archaic instinctive I'.[78] This new, but nevertheless resolutely material, word indeed becomes a tool of a violent, though revealing, penetration. In Harry Crosby's remarkably embodied terms, the 'New Word is the clean piercing of a sword through the rotten carcass of the dictionary'.[79] And the late modernist, neo-Romantic manifesto 'Poetry is Vertical' that appears in *transition* 21 affirms:

The transcendental 'I' with its multiple stratifications reaching back millions of years . . . is brought to the surface with the hallucinatory irruption of images in the dream, the daydream, the mystic-gnostic trance, and even the psychiatric condition.

The final disintegration of the 'I' in the creative act is made possible by the use of language which is a mantic instrument, and which does not hesitate to adopt a revolutionary attitude toward word and syntax.[80]

For Jolas and his co-signatories, language becomes both a site that registers the disintegration of the rational subject, and a ritual 'mantic instrument' – a trepan, of sorts – through which a mystical and prelogical
self and its expressive language might be released.

Jolas implicitly follows psychoanalysis's sense that language can
expose and reconfigure the topography of what lies inside the head,[81]
by suggesting that poetic language is a tool that can penetrate into the
living brain vertically, opening up the cranium to visibility and what is
within to release. In 'Night-Mind and Day-Mind', Jolas affirms:

> We have today means for investigating the night-mind and day-
> mind that never existed before. We must use them and probe as far
> as human possibilities will permit. Modern science is no longer afraid
> of mysticism [. . .] The imagination becomes exact [. . .] Psychology
> has opened the gates of the chthonian world. It is a world within
> our reach.[82]

Here, the 'means', of course, are language. Revealingly, though, alongside an idea of penetrative capacity of the word learned from 'psychology', Jolas remains discursively bound to a sense that getting into
the cranium requires a perforation of matter – a penetrative violence
learned from medicine and war. To read this penetration as a kind of
trepanning is suggestive, then, not simply because it accords with Jolas's
fascination with the primal, the magical, the sacred.[83] Trepanning
becomes metaphorically significant because although its exact purpose
remains unclear, the accounts of both medical and ritual practices tend
to concur that its underlying aim is to create a new orifice in the skull
because the given ways in and out of the cranium become insufficient,
either to let something out – pus, pressure, vapours, spirits; or to let
something in – air, light, inspiration or the gods.[84] The hole bored into
the cranium thus produces an orifice through which something other
than the normative self can speak and be seen. Trepanning functions
here as a wounding which nevertheless promises a certain healing in
the form of a new, mystically reconfigured subject – a subject who has
sloughed off the shape of the man machine.

For Jolas, then, what emerges through and as this orifice metaphorically bored into the cranium is an elision of language and vision that
moves beyond the anatomizing, dissecting gaze of Foucault's clinical
bone-breaker. For the creation of a new orifice in the skull that exposes
the mystical pineal body beneath the 'big brain' does not simply carve

out a new mouth in the head; it also opens up what Jolas describes as a new eye capable of an endopsychic perception that will subtend this new means and mode of expression. Jolas's language allegedly bores open '[t]he inner eye.) [sic] The third eye. The attempt to express an absolute reality with a-logical means'.[85] Jolas indeed articulates the possibility of a unity of vision and language that will be able to perceive and articulate, without distortion, the reality of the 'night mind' which is, for him, the essence of the human personality – 'a language that will dance and sing, that will be the vision of the "troisième oeil", that will bind the races in a fabulous unity'.[86]

By opening up a hole in the skull through which something might enter from above or be released from below, Jolas affirms his mysticism and the new vision and expression of the 'orphic poet' who works ritually to 'penetrate still deeper into the most abstract limits of his conscious and sub-conscious'.[87] But in insisting that it is the skull that must be opened up using the newly roughened-up, hardened and sharpened shapes of material language, in imagining new forms of vision which could penetrate into the brain matter in which mind and language are felt decisively to be held, the pre-modern ritual of trepanning also makes contact with the discourses and obsessions of a technological modernity. For just as Laszlo Moholy-Nagy, who published photographs and essays in *transition*, describes photographic X-ray images as a new form of '[p]enetrative seeing' in 'A New Instrument of Vision' (1934),[88] Jolas too suggests that the technologies of photography and cinema with which *transition* remains obsessed, as it tries to align its twentieth-century word with the machines and expressive possibilities it imagines will reach beyond previous representational realisms,[89] might open up new realms of vision and expression: 'You brought me the gift of the new century, divine affirmer, sirocco of electricity, pulsing of radio, express train, cinema, television! . . . Immensely the world lies before me, x-rayed in every fibre of its cosmic magic,' he affirms.[90] He states clearly in 1929: 'We are not against the machine . . . We may use the mystery of the new instruments as a base from which to proceed into the world of adventures'.[91] But of course, as we have seen and despite Jolas's implied hopes, the X-ray could not see into the functioning brain in a way that revealed anything about thought and language; consequently, Jolas had to imagine another penetrative, pseudo-radiographic, tool with the 'magical' capacity to reveal and reshape brain matter. Neurology had demonstrated that penetrating the live skull could release language and sounds that refused to subtend normative versions of subjectivity, but it had also shown, as a psychoanalytic method emerged from it (and in the face of the limits of X-rays in relation to the brain), that trepanation by the word could open up and see into the cranium and then effect certain forms of repair. So the aesthetic of trepanation in *transition* suggests a language that recognises the shocking sounds and materialized forces that erupt from modernity's particular 'crossroads of matter', but in metaphorically exposing and penetrating the soft living brain using a puncturing, perforating word that reveals the shape of cranial interiority, Jolas uncovers a material, linguistic, speaking and writing subject who is translated away from a 'nineteenth century [that] mechanized the body' and the mind towards a primal past and future.[92] The linguistically trepanned revolutionary of the word paradoxically uses a technological imaginary to speak and write beyond the structures and shapes of the nineteenth-century machine.

Notes

1. John Cournos, 'Henri Gaudier-Brezska', *The Egoist*, 2.8 (1915), 121.
2. Guillaume Apollinaire, 'The New Spirit and the Poets', *Selected Writings of Guillaume Apollinaire*, trans. Roger Shattuck (New York: New Directions, 1971), 232.
3. T.S. Eliot, 'Whispers of Immortality', *Poems* (New York: Alfred A. Knopf, 1920), 31.
4. Apollinaire, 'The New Spirit and the Poets', 232.
5. Paul Broca, 'Notes on the Site of the Faculty of Articulated Language, followed by an Observation of Aphemia' (1861), in Paul Eling (ed.), *Reader in the History of Aphasia: From Franz Gall to Norman Geschwind* (repr., Amsterdam: John Benjamins, 1994), 43.
6. Carl Wernicke, 'The Motor Speech Path and the Relation of Aphasia to Anarthia', in Gertrude H. Eggert (ed.), *Wernicke's Works on Aphasia: A Sourcebook and Review* (1884; repr., The Hague: Mouten, 1977), 152–53.
7. Sigmund Freud, *On Aphasia: A Critical Study* (1890; repr., London: Imago, 1953), 2.
8. John Forrester, *Language and the Origins of Psychoanalysis* (London and Basingstoke: Macmillan, 1980), 16.
9. See Anne Harrington, *Reenchanted Science: Holism in German Culture from Wilhelm II to Hitler* (Princeton: Princeton University Press, 1996), 14.
10. Thomas Carlyle, *Sartor Resartus: The Life and Opinions of Herr Teufelsdröckh* (Berkeley and Los Angeles: University of California Press, 2000), 190.
11. Harrington, *Reenchanted Science*, 14.
12. Thomas Alva Edison, '"No Immortality of the Soul", says Edison', *New York Times* (2 October 1910), 1.
13. Edison, '"No Immortality of the Soul", says Edison', 1.

14. Hermann von Helmholtz, 'On the Conservation of Force', *Science and Culture: Popular and Philosophical Essays*, trans. David Cahan (Chicago: University of Chicago Press, 1995), 98.
15. Harrington, *Reenchanted Science*, 8.
16. Rudolf Virchow, 'On the Mechanistic Interpretation of Life', *Disease, Life and Man: Selected Essays by Rudolf Virchow*, trans. Lelland J. Rather (Stanford: Stanford University Press, 1958), 107.
17. Edwin Clarke and L.S. Jacyna, *Nineteenth-Century Origins of Neuroscientific Concepts* (Berkeley and Los Angeles: University of California Press, 1987), 220–44.
18. Michel Foucault, *The Birth of the Clinic*, trans. A.M. Sheridan (London: Tavistock Publications, 1976), 141.
19. Foucault, *The Birth of the Clinic*, xiii.
20. Foucault, *The Birth of the Clinic*, xiii.
21. Foucault, *The Birth of the Clinic*, xiv.
22. Foucault, *The Birth of the Clinic*, 131.
23. Foucault, *The Birth of the Clinic*, 131.
24. Foucault, *The Birth of the Clinic*, xix.
25. Foucault, *The Birth of the Clinic*, 131.
26. Of course, X-rays did penetrate the body, causing their own very particular forms of damage. As Lisa Cartwright notes, by 1903 Edison had revealed to and through the newspapers that X-rays could cause injuries, and as early as 1904, his assistant Clarence Dally had died from X-ray-induced carcinoma (*Screening the Body: Tracing Medicine's Visual Culture* [Minneapolis: University of Minnesota Press, 1995], 109–10).
27. Cited in Cartwright, *Screening the Body*, 183, n.13.
28. X-rays could be used to detect foreign objects in the brain, as in Apollinaire's case, and two techniques for imaging tumours and malformations emerged in the 1910s and 1920s which involved penetrating the body and inserting air or contrast substances into the brain which could then be mapped radiographically (J.W.D. Bull, 'The History of Neuroradiology', in F. Clifford Rose and W.F. Bynum (eds), *Historical Aspects of Neuroscience* [New York: Raven Press, 1982], 257–59).
29. William Shuler Harris, *Life in a Thousand Worlds* (New York: Arno Press, 1971), 236.
30. Marcus E. Raichle and Mark A. Mintun, 'Brain Work and Brain Imaging', *Annual Review of Neuroscience*, 29 (2006), 449–76.
31. Freud, *On Aphasia*, 62.
32. Freud, *On Aphasia*, 62.
33. Freud, *On Aphasia*, 62.
34. Freud, *On Aphasia*, 51.
35. Freud, *On Aphasia*, 53.
36. Forrester, *Language and the Origins of Psychoanalysis*, 24.
37. Freud, *On Aphasia*, 103.
38. Freud, *On Aphasia*, 67.
39. Freud, *On Aphasia*, 30.
40. L.S. Jacyna, *Lost Words: Narratives of Language and the Brain, 1825–1926* (Princeton: Princeton University Press, 2000), 181.

41. Valerie Greenberg reads a 'conservatism' in those few diagrams that remain in Freud's text because they seem to preserve theories of brain function 'in fixed lines, as opposed to the fluidity of meaning in shifting and pliable language' (*Freud and His Aphasia Book: Language and the Sources of Psychoanalysis* [Ithaca and London: Cornell University Press, 1997], 168).

42. Freud was to remain conflictedly attached to notions of depth and visual penetration. Mark Solms has noted that, alongside those famous archaeological metaphors that structure Freud's accounts of psychoanalytic theory and his spatial topography of the mind, his diagrams of psychological functioning are remarkably similar in their structure to the neurological diagrams he uses to explain aphasia (Lynn Gamwell and Mark Solms, *From Neurology to Psychoanalysis: Sigmund Freud's Neurological Drawing and Diagrams of the Mind* [New York: SUNY, 2006], 16–17). For the theoretical Freud, the verbal and the temporal was never fully to transcend the visual and the spatial.

43. Sander L. Gilman, *Seeing the Insane* (New York: John Wiley, 1982), 223.

44. Sigmund Freud, *Studies on Hysteria, The Standard Edition of the Complete Psychological Works of Sigmund Freud*, vol. 2 (1893–95; repr., London: Hogarth Press, 1955), 29, my emphasis.

45. Sigmund Freud, *The Complete Letters of Sigmund Freud to Wilhelm Fliess: 1887–1904*, trans. and ed. Jeffrey Moussaieff Masson (Cambridge, MA: Harvard University Press, 1985), 207.

46. Forrester, *Language and the Origins of Psychoanalysis*, 30.

47. Samuel H. Greenblatt, 'Cerebral Localization: From Theory to Practice', in Samuel H. Greenblatt, T. Forcht Dagi and Mel H. Epstein (eds), *A History of Neurosurgery in Its Scientific and Professional Contexts* (Park Ridge: American Association of Neurological Surgeons, 1997), 137–52. See also Jacyna's account of William Macewen, a surgeon at the Glasgow Royal Infirmary between 1877 and 1892, who acknowledged that his pioneering brain surgery was built on the studies of Broca and his followers (*Lost Words*, 212–14).

48. In the 1910s, Russian surgeon Ludwig M. Pussep detailed the fact that some aphasic symptoms were indeed relieved by brain surgery. Unsurprisingly, though, general neurosurgical literature from the period suggested that many aphasias remained stubbornly resistant to surgical intervention (Jacyna, *Lost Words*, 212–14).

49. Sigmund Freud, 'Psychical (or Mental) Treatment', in *The Standard Edition of the Complete Psychological Works of Sigmund Freud*, vol. 7 (1890; repr., London: Hogarth Press, 1953), 283.

50. Eugene Jolas, 'Mantic News', *transition*, 21 (1932), 240.

51. Jolas, 'Mantic News', 240.

52. Dougald McMillan, *transition: The History of a Literary Era, 1927–1938* (London: Calder and Boyars, 1975), 1.

53. Eugene Jolas, 'Super-Occident', *transition*, 15 (1929), 15.

54. Eugene Jolas, 'Proclamation: The Revolution of the Word', *transition*, 16/17 (1929), 13.

55. Eugene Jolas, *The Language of the Night* (The Hague: Servire Press, 1932), 40.

56. Jolas, *The Language of the Night*, 60.

57. Jolas, *The Language of the Night*, 66.

58. Eugene Jolas, *Man from Babel* (New Haven and London: Yale University Press, 1998), 272.
59. Jolas, *Man From Babel*, 2.
60. Jolas, *Man From Babel*, 53.
61. Vincent Sherry, *The Great War and the Language of Modernism* (Oxford: Oxford University Press, 2003), 16.
62. Eugene Jolas et al, 'Poetry is Vertical', *transition*, 21 (1932), 148. For an account of aphasiology's own refusal of mechanized models of brain function that emerged directly from the Great War, see Laura Salisbury, 'Sounds of Silence: Aphasiology and the Subject of Modernity', in Laura Salisbury and Andrew Shail (eds), *Neurology and Modernity* (Basingstoke: Palgrave Macmillan, 2010), 204–30. See also Jacyna, *Lost Words*, 146–70, 225–30.
63. Jolas, *The Language of the Night*, 40.
64. Jolas, *The Language of the Night*, 43.
65. Jolas, *The Language of the Night*, 41.
66. John Cournos, 'Henri Gaudier-Brezska', 121.
67. Jolas, *Man From Babel*, 36.
68. Eugene Jolas, 'Notes on Reality', *transition*, 18 (1929), 16.
69. Stuart Gilbert, 'Art and Intuition', *transition*, 21 (1932), 216.
70. Gilbert, 'Art and Intuition', 220–21.
71. For an account of the relationship between *transition*'s anti-mechanistic models of brain function and neurology's postwar refusal of the idea of the cerebral machine that emerged from Gestalt psychology, see Laura Salisbury, '"What Is the Word": Beckett's Aphasic Modernism', *Journal of Beckett Studies*, 17 (2008), 97–115.
72. Gottfried Benn, 'The Structure of Personality', *transition*, 21 (1932), 196, 199.
73. Benn, 'The Structure of Personality', 196, 199.
74. Jolas, *The Language of the Night*, 48.
75. Benn, 'The Structure of Personality', 200.
76. 'Jackson, in connection with Spencer's evolutionary theory, spoke of the dissolution of complicated functional systems in psychosis by regression to a lower scale' (Benn, 'The Structure of Personality', 201). See John Hughlings Jackson, 'On Affections of Speech from Disease of the Brain', in Paul Eling (ed.), *Reader in the History of Aphasia: From Franz Gall to Norman Geschwind* (Amsterdam: John Benjamins, 1994), 149.
77. Jolas et al., 'Poetry is Vertical', 148.
78. Benn, 'The Structure of Personality', 200.
79. Harry Crosby, 'The New Word', *transition*, 16/17 (1929), 30.
80. Jolas et al., 'Poetry is Vertical', 148.
81. Although *transition* remains more sympathetically aligned with the psycho-analytic theories of Jung than Freud (see Jolas, 'Notes on Reality', 16–17), Jolas insists on the importance of psychoanalysis as a theory and practice for the aims of the 'Revolution of the Word'.
82. Eugene Jolas, 'Night-Mind and Day-Mind', *transition*, 21 (1932), 223.
83. Jolas, like many of the other members of the avant-garde, was fascinated by primitivism and anthropology, and in *transition* 19/20 he reproduced two pictures of ritual skulls allegedly used by pre-modern peoples (377, 380).

84. Edward L. Margetts, 'Trepanation of the Skull by the Medicine Men of Primitive Cultures, with particular reference to present-day East-African practice', in Donald R. Brothwell and A.T. Sandison (eds), *Diseases in Antiquity* (Springfield: Charles C. Thomas, 1967), 673–701.
85. Eugene Jolas, 'The Primal Personality', *transition*, 22 (1933), 82.
86. Eugene Jolas, 'Workshop', *transition*, 23 (1935), 104.
87. Eugene Jolas, 'Logos', *transition*, 16/17 (1929), 26.
88. Laszlo Moholy-Nagy, 'A New Instrument of Vision', in Liz Wells (ed.), *The Photography Reader* (London: Routledge, 2003), 94.
89. See Michael North, *Camera-Works: Photography and the Twentieth-Century Word* (Oxford: Oxford University Press, 2005), 61–82.
90. Eugene Jolas, 'Construction of the Enigma', *transition*, 15 (1929), 60.
91. Eugene Jolas, 'Preface', in *Transition Stories: Twenty-three Stories from 'Transition'*, ed. Eugene Jolas and Robert Sage (New York: Walter V. McKee, 1929), p. x.
92. Jolas, 'The Primal Personality', 80.

Coda: From the posthumous to the posthuman

The Visible Human Project, sponsored by the US National Library of Medicine, has brought anatomy into the digital age. Accessible to all with a computer and internet connection are two visible humans, a Visible Human Female and a Visible Human Man, both of whom were once living, breathing people. The woman's identity has never been revealed. All we know is that she was fifty-nine years old when she died of a heart attack in Maryland, USA. She donated her body to medical science. The man's name is Joseph Jernigan, a thirty-nine-year-old death-row inmate who, at the urging of the prison chaplain, consented to donate his body after death by lethal injection. Both bodies were embalmed, frozen, cut into four parts, and then cross-sectioned into microscopic slices using three technologies: magnetic resonance imaging, computerized tomography, and digital photography. Volumetric rendering was applied to give the data-sets an appearance of three-dimensional volume and depth, and finally, the images were animated. Some of these images can be viewed at www.nlm.nih.gov/research/visible/visible_human.html

The purpose of the Visible Human Project was to create a digital archive of the human body that could be put to both commercial and academic use. For the casual visitor the website opens up a number of possibilities. You can devise your own animated trip through the cryosections or you can to go on a guided tour called *Body Voyage: A Three-Dimensional Tour of a Real Human Body*. For those with more serious interests the site offers interactive exploration of cross-sectional anatomy in pixel-level detail. That these once-living bodies now enjoy a posthumous existence as digitized information gives a whole new meaning to Kevin Warwick's downloading of himself into a computer. But while Warwick's object appears to be the realization of immortality through technological mastery, it is the human vulnerability to technology which is the most striking feature of this project. In their transformation into pixels these people have become 'things', visual data-sets on computer screens with no interiority or privacy. If the interior of the body has traditionally been seen as the domain of the soul, then the destruction of all interiority – and allied with this, a destruction of personhood – causes some disquiet in viewers. Questions concerning the boundaries between the dead body and the living person are particularly pressing, as noted by anthropologist Rachel Prentice, who interviewed several visitors to the Visible Human Project site. Also prominent are ethical issues around consent. If these people had known, when alive, that they were to be visible for eternity, would they have consented to such an afterlife?[1] Could they ever have imagined their bodies' dismemberment and subjection to new orders of machine vision, or their eerie reanimation in a virtual space?[2]

The unnerving proximity of life and death, and the mediation of this proximity by technology, recall Derrida's reflections on the iterability and hence the spectrality of the photograph which join, in a single system, death and its referent. For Cary Wolfe, in his new book, *What is Posthumanism* (2010), Derrida's posthumanism 'finds the generative force of the nonliving at the origins of any living being, human *or* animal, who communicates . . . with another'. Signalled here at the heart of Wolfe's book is the currently burgeoning field of interdisciplinary enquiry called human-animal studies. In a chapter on bioethics entitled 'Flesh and Finitude', Wolfe critiques the analytic tradition of philosophy for the ungroundedness of concepts such as 'person', and its circumscribed understanding of the complex questions of life, death and our relation to other living beings. Traditional philosophy, even in its most sensitive approach to issues of anthropocentrism and speciesism, fails to square up to the challenge of 'sharing the planet with non-human subjects and treating them justly'. Rejecting the central role of rationality as the chief ontological and ethical dividing line between humans and nonhuman animals, Wolfe emphasizes that it is our shared embodiment – mortality and finitude – that makes us fellow-creatures.

Human/Animal boundaries are at the heart of Wolfe's book but his definition of posthumanism extends outwards to the boundaries between the organic and the technological. Instead of identifying the posthuman with fantasies of a triumphant disembodiment and autonomy, Wolfe defines the posthuman in opposition to such fantasies, which he sees as the invidious legacy of humanism. His definition has two points of emphasis, one coming before humanism, the other after it. Before humanism entails 'the embodiment and embeddedness of the human being in not just its biological but also its technological world, the prosthetic co-evolution of the human animal with the technicity of tools and archival mechanisms (such as language and culture)'. The 'after' of Wolfe's posthumanism names our historical present, in which it is increasingly impossible to ignore 'the decentering of the human by its imbrication in technical, medical, informatic, and economic networks'. In addition to our shared vulnerability as embodied creatures there is our subjection to language and writing, with our relation to flesh and blood, in the words of Derrida, 'fatefully constituted by a technicity with which it is prosthetically entwined, a diacritical, semiotic machine of language'. Here, in Wolfe's centring of posthumanism in the criss-crossing of animal-human studies with explorations of organism-machine boundaries, we move out beyond present preoccupations with transhumanism – our bodily enhancement through prostheses, robotics, computers, and virtual reality system – into new cross-disciplinary fields of inquiry.[3]

Notes

1. Rachel Prentice, 'The Visible Human' in Sherry Turkle (ed.), *The Inner History of Devices* (Cambridge, MA and London: MIT Press, 2008), 112–24.
2. See Catherine Waldby, *The Visible Human Project: Informatic Bodies and Posthuman Medicine* (London and New York: Routledge, 2000), 5–6.
3. Cary Wolfe, *What is Posthumanism?* (Minneapolis and London: University of Minnesota Press, 2010), 62.

Bibliography

Aikin, John and Aikin, Anna Letitia, 'On the Pleasure derived from Objects of Terror; with Sir Bertrand, a Fragment', in *Miscellaneous Pieces, in Prose* (London, 1773): 117–37.

Aldini, John, *General Views on the Application of Galvanism to Medical Purposes; Principally in Cases of Suspended Animation* (London: Callow, 1819).

Anon, 'Austro-Hungary', *The Times* (10 December 1889): 5.

Anon, 'THE COMING EPIDEMIC! THE COMING EPIDEMIC!!!', *Illustrated London News*, 96 (1890): 31.

Anon, 'The Epidemic of Influenza', *The Times* (17 December 1889): 5.

Anon, 'The Epidemic of Influenza', *The Times* (18 December 1889): 5.

Anon, 'Foreign and Colonial News', *The Times* (14 December 1889): 5.

Anon, 'Germany', *The Times* (9 December 1889): 6.

Anon, 'Has Influenza: How to Cure It and How to Prevent It', *Pall Mall Gazette* (9 January 1890): 1.

Anon, 'The Influenza', *Spectator*, 66 (1891): 718–19.

Anon, 'The Influenza Epidemic', *Pall Mall Gazette* (9 December 1889): 4.

Anon, 'The Influenza Epidemic at Home and Abroad', *Pall Mall Gazette* (11 December 1889): 4.

Anon, 'The Influenza Epidemic in Russia', *Lancet*, 134 (1889): 1134.

Anon, 'London: Friday, December 13, 1889', *The Times* (13 December 1889): 9.

Anon, 'London: Saturday, December 21, 1889', *Lancet*, 134 (1889): 1293–94.

Anon, 'London: Tuesday, December 3, 1889', *The Times* (3 December 1889): 9.

Anon, 'Our London Letter', *Northern Echo* (11 December 1889): 3.

Anon, 'Russia', *The Times* (25 November 1889): 6.

Anon, 'Russia', *The Times* (30 November 1889): 5.

Anon, 'This Morning's News', *The Leeds Mercury* (9 December 1889): 5.

Anon, 'To-Day's Tittle-Tattle', *Pall Mall Gazette* (1 January 1890): 5.

Anon, 'The users of SALT REGAL have hitherto escaped THE EPIDEMIC', *Illustrated London News*, 96 (1890): 96.

Apollinaire, Guillaume, 'The New Spirit and the Poets', in *Selected Writings of Guillaume Apollinaire*, trans. Roger Shattuck (New York: New Directions, 1971): 227–37.

Armstrong, Nancy and Tennenhouse, Leonard, 'A Mind for Passion: Locke and Hutcheson on Desire', in Victoria Kahn, Neil Saccamano and Daniela Coli, eds, *Politics and the Passions, 1500–1850* (Princeton: Princeton University Press, 2006): 131–50.

Ashton, Rosemary, *G. H. Lewes: An Unconventional Victorian* (London: Pimlico, 2000).

Bain, Alexander, 'Common Errors on the Mind', *Fortnightly Review*, New Series, 4 (1868): 160–75.

Bain, Alexander, 'The Feelings and the Will Viewed Physiologically', *Fortnightly Review*, 3 (1866): 575–78.

Bain, Alexander, 'On the Correlation of Force in its Bearing on the Mind', *Macmillan's Magazine*, 16 (1867): 372–83.

Bain, Alexander, *The Senses and the Intellect* (London: John W. Parker, 1855).

Barrell, John, 'Putting Down the Rising', in Leith Davis, Ian Duncan and Janet Sorensen, eds, *Scotland and the Borders of Romanticism* (Cambridge: Cambridge University Press, 2004): 130–8.

Barthes, Roland, 'The Death of the Author', in *Image-Music-Text* (London: Fontana, 1977): 142–8.

Bates, David, 'Creating Insight: Gestalt Theory and the Early Computer', in Jessica Riskin, ed., *Genesis Redux: Essays in the History and Philosophy of Artificial Life* (Chicago: Chicago University Press, 2007): 237–62.

Batchen, Geoffrey, *Burning with Desire: The Conception of Photography* (Cambridge, MA: MIT Press, 1997).

Beer, Gillian, *Meredith: A Change of Masks. A Study of the Novels* (London: Athlone, 1970).

Beer, Gillian, *Open Fields: Science in Cultural Encounter* (Oxford: Clarendon Press, 1996).

Bender, John, *Imagining the Penitentiary: Fiction and the Architecture of Mind in Eighteenth-Century England* (Chicago: University of Chicago Press, 1987).

Benn, Gottfried, 'The Structure of Personality', *transition*, 21 (1932): 195–205.

Bennett, Andrew, *The Author* (London: Routledge, 2005).

Blackmore, Susan, *Consciousness: An Introduction* (London: Hodder and Stoughton, 2003).

Blackwell, Bonnie, '*Tristram Shandy* and the Theater of the Mechanical Mother', *ELH*, 68 (2001): 81–133.

Bloedé, Barbara, 'Hogg and the Edinburgh Prostitution Scandal', *Newsletter of the James Hogg Society*, 8 (1989): 15–18.

Bloedé, Barbara, '*The Three Perils of Woman* and the Edinburgh Prostitution Scandal: A Reply to Dr Groves', *Studies in Hogg and His World*, 3 (1992): 88–94.

Bloom, Harold, *The Anxiety of Influence: A Theory of Poetry* (Oxford: Oxford University Press, 1973).

Bök, Christian, 'The Piecemeal Bard is Deconstructed: Notes Toward a Potential Robopoetics', *Object*, 10 (2002): 10–18.

Brantlinger, Patrick, 'Mass Media and Culture in Late Nineteenth-Century Europe', in Mikuláš Teich and Roy Porter, eds, *Fin de Siècle and its Legacy* (Cambridge: Cambridge University Press, 1990): 98–114.

Brewster, David, *Edinburgh Encyclopaedia* (1830; repr. London: Routledge, 1999).

Briggs, Asa, *Victorian Things*, 3rd edition (Stroud: Sutton, 2003).

Bringsjord, Selmer and Ferrucci, David A., *Artificial Intelligence and Literary Creativity: Inside the mind of BRUTUS, A Storytelling Machine* (Mahwah: Lawrence Erlbaum, 2000).

Broca, Paul, 'Notes on the Site of the Faculty of Articulated Language, followed by an Observation of Aphemia' (1861), in Paul Eling, ed., *Reader in the History of Aphasia: From Franz Gall to Norman Geschwind* (Amsterdam: John Benjamins, 1994): 41–45.

Brock, Thomas D., *Robert Koch: A Life in Medicine and Bacteriology* (Madison, WI: SciTech Press, 1988).

Broks, Peter, *Media Science Before the Great War* (London: Macmillan, 1996).

Brown, Bill, 'Thing Theory', *Critical Inquiry*, 28 (2001): 1–21.

Brown, Samuel, 'Physical Puritanism', *Westminster Review*, 112 (1852): 405–42.

Bull, J.W.D., 'The History of Neuroradiology', in F. Clifford Rose and W.F. Bynum, eds, *Historical Aspects of Neuroscience* (New York: Raven Press, 1982): 255–64.

Burke, Seán, *The Death and Return of the Author: Criticism and Subjectivity in Barthes, Foucault, and Derrida* (Edinburgh: Edinburgh University Press, 1992).

Campbell, Lewis and Garnett, William, *The Life of James Clerk Maxwell* (London: Macmillan, 1882).

Carlyle, Thomas, *Sartor Resartus: The Life and Opinions of Herr Teufelsdröckh* (Berkeley and Los Angeles: University of California Press, 2000).

Cartwright, Lisa, *Screening the Body: Tracing Medicine's Visual Culture* (Minneapolis: University of Minnesota Press, 1995).

Cicero, M. Tullius, *Of the Nature of the Gods* (London: Franklin R., 1741).

Clarke, Edwin and Jacyna, L.S., *Nineteenth-Century Origins of Neuroscientific Concepts* (Berkeley and Los Angeles: University of California Press, 1987).

Cline, C.L. (ed.), *The Letters of George Meredith* (Oxford: Clarendon, 1970).

Cody, Lisa Forman, *Birthing the Nation: Sex, Science, and the Conception of Eighteenth-Century Britons* (Oxford: Oxford University Press, 2005).

Coleman, Deirdre and Fraser, Hilary (eds), 'Minds, Bodies, Machines', Special issue, 19: *Interdisciplinary Studies in the Long Nineteenth Century*, 7 (2008).

Coleridge, Samuel Taylor, *Biographia Literaria*, J. Engell and W. Jackson Bate, eds, vol. vii of *The Collected Works of Samuel Taylor Coleridge*, 16 vols, Bollingen Series, 75 (London: Routledge and Kegan Paul; Princeton: Princeton University Press, 1983).

Coleridge, Samuel Taylor, *Lectures, 1808–1819, on Literature*, ed. R.A. Folkes, vol. v of *The Collected Works of Samuel Taylor Coleridge*, 16 vols, Bollingen Series, 75 (London: Routledge and Kegan Paul; Princeton: Princeton University Press, 1983).

Collard, Patrick, *The Development of Microbiology* (Cambridge: Cambridge University Press, 1976).

Connor, Steven, *Dumbstruck: A Cultural History of Ventriloquism* (Oxford: Oxford University Press, 2000).

Cooter, Roger, *The Cultural Meaning of Popular Science: Phrenology and the Organization of Consent in Nineteenth-Century Britain* (Cambridge, London and New York: Cambridge University Press, 1984).

Cournos, John, 'Henri Gaudier-Brezska', *The Egoist*, 2.8 (1915): 121.

Cowper, William, *The Task: A Poem* (London: James Nisbet, 1878).

Crary, Jonathan, 'Modernizing Vision', in Hal Foster, ed., *Vision and Visuality*, Dia Art Foundation Discussions in Contemporary Culture, No. 2 (New York: The New Press, 1988): 29–51.

Crary, Jonathan, *Techniques of the Observer: On Vision and Modernity in the Nineteenth Century* (Cambridge, MA: MIT Press, 1990).

Crosby, Harry, 'The New Word', *transition*, 16/17 (1929): 30.

Cunningham, Allan, *A Biographical and Critical History of the British Literature of the Last Fifty Years* (Paris: Baudry's Foreign Library, 1834).

Darwin, Erasmus, *The Loves of the Plants* (Oxford: Woodstock Books, 1991).

Dasenbrock, Reed Way, *Truth and Consequences: Intentions, Conventions and the New Thematics* (University Park: Pennsylvania State University Press, 2001).

Davis, Michael, *George Eliot and Nineteenth-Century Psychology: Exploring the Unmapped Country* (Aldershot: Ashgate, 2006).

Davis-Floyd, Robbie E., 'The Technocratic Model of Birth', in Philip K. Wilson, ed., *The Medicalization of Obstetrics: Personnel, Practice and Instruments* (New York and London: Garland, 1996): 247–76.

Dawson, Gowan, *Darwin, Literature and Victorian Respectability* (Cambridge: Cambridge University Press, 2007).

Day, Aidan, *Romanticism* (London: Routledge, 1996).

de Saint Fond, B. Faujas, *A Journey through England and Scotland to the Hebrides in 1784*, 2 vols (Glasgow: Hugh Hopkins, 1907).

De Landa, Manuel, *War in the Age of Intelligent Machines* (New York: Zone, 1991).

Deleuze, Gilles, *The Logic of Sense*, trans. Mark Lester (London: Athlone, 1990).

Deleuze, Gilles, *Pure Immanence: Essays on a Life*, trans. Anne Boyman (New York: Zone, 2001).

Descartes, René, *The Philosophical Writings of Descartes*, 3 vols, trans. John Cottingham, Robert Stoothoff and Dugald Murdoch (Cambridge: Cambridge University Press, 1984–1991).

Dibdin, Charles, *Mirth and Metre, consisting of Poems, Serious, Humorous, and Satirical; Songs, Sonnets, Ballads & Bagatelles* (London: Vernor, Hood, and Sharpe, 1807).

Dickens, Charles, *Little Dorrit* (London: Penguin, 1967).

Dixey, F.A., *Epidemic Influenza* (Oxford: Clarendon Press, 1892).

Doane, Mary Ann, *The Emergence of Cinematic Time* (Cambridge, MA: Harvard University Press, 2002).

Dobson, Jessie, 'Some Eighteenth-Century Experiments in Embalming', *Journal of the History of Medicine*, 8 (October 1953): 431–41.

Duncan, Ian, *Scott's Shadow: The Novel in Romantic Edinburgh* (Princeton: Princeton University Press, 2007).

Durant, John R., 'Scientific Naturalism and Social Reform in the Thought of Alfred Russel Wallace', *British Journal for the History of Science*, 12.1 (1979): 31–58.

During, Simon, *Modern Enchantments: The Cultural Power of Secular Magic* (Cambridge, MA: Harvard University Press, 2002).

Edison, Thomas Alva, '"No Immortality of the Soul", says Edison', *New York Times* (2 October 1910): 1.

Eliot, George, *Middlemarch* (Oxford: Oxford University Press, 2008).

Eliot, T.S., *Poems* (New York: Alfred A. Knopf, 1920).

Ernst, Josef, 'Computer Poetry: An Act of Disinterested Communication', *New Literary History*, 23.2 (1992): 451–65.

Ferguson, Frances, *Pornography, the Theory: What Utilitarianism Did to Action* (Chicago and London: University of Chicago Press, 2004).

Forrester, John, *Language and the Origins of Psychoanalysis* (London and Basingstoke: Macmillan, 1980).

Foucault, Michel, *The Birth of the Clinic*, trans. A.M. Sheridan (London: Tavistock Publications, 1976).

Freedgood, Elaine, *The Ideas in Things: Fugitive Meanings in the Victorian Novel* (Chicago: University of Chicago Press, 2006).

Freud, Sigmund, *The Complete Letters of Sigmund Freud to Wilhelm Fliess: 1887–1904*, ed. and trans. Jeffrey Moussaieff Masson (Cambridge, MA: Harvard University Press, 1985).

Freud, Sigmund, *On Aphasia: A Critical Study* (1890; repr., London: Imago, 1953).

Freud, Sigmund, 'Psychical (or Mental) Treatment', in *The Standard Edition of the Complete Psychological Works of Sigmund Freud*, vol. 7 (1890; repr., London: Hogarth Press, 1953).

Freud, Sigmund, *Studies on Hysteria, The Standard Edition of the Complete Psychological Works of Sigmund Freud*, vol. 2 (1893–95; repr., London: Hogarth Press, 1955).

Freud, Sigmund, 'The Uncanny', in *Art and Literature*, ed. Albert Dickson (London: Penguin, 1990).

Gamwell, Lynn and Solms, Mark, *From Neurology to Psychoanalysis: Sigmund Freud's Neurological Drawing and Diagrams of the Mind* (New York: SUNY, 2006).

Gilbert, Stuart, 'Art and Intuition', *transition*, 21 (1932): 215–21.

Gilman, Sander L., *Seeing the Insane* (New York: John Wiley, 1982).

Glaister, John, *Dr. William Smellie and his Contemporaries: A Contribution to the History of Midwifery in the Eighteenth Century* (Glasgow: Maclehose, 1894).

Greenberg, Valerie, *Freud and His Aphasia Book: Language and the Sources of Psychoanalysis* (Ithaca and London: Cornell University Press, 1997).

Greenblatt, Samuel H., 'Cerebral Localization: From Theory to Practice', in Samuel H. Greenblatt, T. Forcht Dagi and Mel H. Epstein, eds, *A History of Neurosurgery in its Scientific and Professional Contexts* (Park Ridge: American Association of Neurological Surgeons, 1997): 137–52.

Groves, David, 'Afterword', in *The Three Perils of a Woman, or Love, Leasing, and Jealousy, a series of Domestic Scottish Tales*, by James Hogg (Edinburgh: Edinburgh University Press, 1995): 409–37.

Groves, David, 'James Hogg's *Confessions* and *The Three Perils of Woman* and the Edinburgh Prostitution Scandal of 1823', *Wordsworth Circle*, 18 (1987): 127–31.

Groves, David, '*The Three Perils of Woman* and the Edinburgh Prostitution Scandal of 1823', *Studies in Hogg and his World*, 2 (1991): 95–102.

Haraway, Donna, *Simians, Cyborgs and Women: The Reinvention of Nature* (New York: Routledge, 1991).

Harman, P.M. (ed.), *The Scientific Letters and Papers of James Clerk Maxwell* (Cambridge: Cambridge University Press, 1990–2001).

Harrington, Anne, *Reenchanted Science: Holism in German Culture from Wilhelm II to Hitler* (Princeton: Princeton University Press, 1996).

Harris, William Shuler, *Life in a Thousand Worlds* (New York: Arno Press, 1971).

Harrison, J.F.C., *Quest for the New Moral World. Robert Owen and the Owenites in England and America* (New York: Scribner, 1969).

Harrison, Ross, *Bentham* (London: Routledge and Kegan Paul, 1983).

Haslam, John, *Illustrations of Madness* (1810; repr., London and New York: Routledge, 1988).

Haslam, John, *Observations on Madness and Melancholy: Including Practical Remarks on Those Diseases; Together With Cases: And An Account of the Morbid Appearances on Dissection*, 2nd edition (London: for J. Callow, 1809).

Hasler, Antony J., 'Introduction', in *The Three Perils of Woman, or Love, Leasing, and Jealousy, a series of Domestic Scottish Tales*, by James Hogg (Edinburgh: Edinburgh University Press, 2002).

Hasler, Antony J., 'The Three Perils of Woman and John Wilson's *Lights and Shadows of Scottish Life*', *Studies in Hogg and his World*, 1 (1990): 30–45.

Hayles, N. Katherine, *Electronic Literature: New Horizons for the Literary* (Notre Dame: University of Notre Dame Press, 2008).

Hayles, N. Katherine, *How We Became Posthuman: Virtual Bodies in Cybernetics, Literature, and Informatics* (Chicago: University of Chicago Press, 1999).

Hayles, N. Katherine, 'The Materiality of Informatics', *Configurations*, 1.1 (1993): 147–70.

Hayles, N. Katherine, 'Virtual Bodies and Flickering Signifiers', in Timothy Druckery, ed., *Electronic Culture* (New York: Aperture, 1996): 259–77.

Helmreich, Stefan, 'Flexible Infections: Computer Viruses, Human Bodies, Nation-States, Evolutionary Capitalism', *Science, Technology and Human Values*, 25.4 (2000): 472–91.

Henderson, Andrea K., *Romantic Identities: Varieties of Subjectivity 1774–1830* (Cambridge: Cambridge University Press, 1996).

Hibbard, Bryan, *The Obstetrician's Armamentarium: Historical Obstetric Instruments and Their Inventors* (San Anselmo: Norman, 2000).

Hobbes, Thomas, *The Leviathan* (Toronto: Broadview Literary Texts, 2002).

Hoffmann, E.T.A., 'Automata', in *The Best Tales of Hoffmann*, ed. E.F. Bleiler, trans. Major Alexander Ewing (Mineola: Dover, 1967): 71–103.

Hoffmann, E.T.A., 'The Sandman', in *Tales from the German, Comprising Specimens from the Most Celebrated Authors*, trans. John Oxenford and C.A. Feiling (London: Chapman and Hall, 1844): 140–65.

Hogg, James, *Anecdotes of Scott* (Edinburgh: Edinburgh University Press, 2004).

Hogg, James, *The Three Perils of Woman, or Love, Leasing, and Jealousy, a series of Domestic Scottish Tales* (Edinburgh: Edinburgh University Press, 2002).

Hughes, Gillian, *James Hogg: A Life* (Edinburgh: Edinburgh University Press, 2007).

Hume, David, *A Treatise of Human Nature*, ed. Ernest C. Mossner (London: Penguin, 1985).

Hunter, William, *The Anatomy of the Gravid Uterus Exhibited in Figures* (Birmingham: Baskerville, 1774).

Jackson, John Hughlings, 'On Affections of Speech from Disease of the Brain', in Paul Eling, ed., *Reader in the History of Aphasia: From Franz Gall to Norman Geschwind* (Amsterdam: John Benjamins, 1994): 145–67.

Jackson, Richard D., 'Gatty Bell's Illness in James Hogg's *The Three Perils of Woman*', *Studies in Hogg and His World*, 14 (2003): 16–29.

Jacyna, L.S., *Lost Words: Narratives of Language and the Brain, 1825–1926* (Princeton: Princeton University Press, 2000).

James, Henry, *The Portrait of a Lady* (Oxford: Oxford University Press, 2008).

James, William, *Writings, 1878–1899*, ed. Gerald E. Myers (New York: Library of America, 1992).

Jay, Mike, *The Air-Loom Gang; The Strange and True Story of James Tilly Matthews and His Visionary Madness* (London and New York: Bantam Books, 2004).

Jefferson, Geoffrey, 'The Mind of Mechanical Man', *British Medical Journal*, 1.4616 (25 June 1949): 1105–10.

Jenkin, Fleeming, 'The Atomic Theory of Lucretius', *North British Review*, 48 (1868): 211–42.

Johnson, Galen A. (ed. and trans.), *The Merleau-Ponty Aesthetics Reader: Philosophy and Painting* (Evanston: Northwestern University Press, 1993).

Johnson, Michael L., *Mind, Language, Machine: Artificial Intelligence in the Poststructuralist Age* (Basingstoke: Macmillan, 1988).

Johnstone, R.W., *William Smellie: The Master of British Midwifery* (Edinburgh: E&S Livingstone, 1952).

Jolas, Eugene, 'Construction of the Enigma', *transition*, 15 (1929): 56–62.

Jolas, Eugene, 'GIRL, OPERATED ON FOR AMNESIA, SPEAKS 12 LANGUAGES', *transition*, 21 (1932): 240.

Jolas, Eugene, *The Language of the Night* (The Hague: Servire Press, 1932).

Jolas, Eugene, 'Logos', *transition*, 16/17 (1929): 25–29.

Jolas, Eugene, *Man from Babel* (New Haven and London: Yale University Press, 1998).

Jolas, Eugene, 'Night-Mind and Day-Mind', *transition*, 21 (1932): 222–23.

Jolas, Eugene, 'Notes on Reality', *transition*, 18 (1929): 15–21.

Jolas, Eugene, et al., 'Poetry is Vertical', *transition*, 21 (1932): 148–49.

Jolas, Eugene, 'Preface', *transition*, 21 (1932): 6.

Jolas, Eugene, 'The Primal Personality', *transition*, 22 (1933): 78–83.

Jolas, Eugene, 'Proclamation: The Revolution of the Word', *transition*, 16/17 (1929): 13.

Jolas, Eugene, 'Super-Occident', *transition*, 15 (1929): 11–16.

Jolas, Eugene, 'Workshop', *transition*, 23 (1935): 97–106.

Jones, Greta, 'Alfred Russel Wallace, Robert Owen and the Theory of Natural Selection', *BJHS*, 35 (2002): 73–96.

Jordanova, Ludmilla, *Sexual Visions: Images of Gender in Science and Medicine between the Eighteenth and Twentieth Centuries* (Hemel Hempstead: Harvester Wheatsheaf, 1989).

Keep, Christopher, 'Of Writing Machines and Scholar-Gypsies', *English Studies in Canada*, 29.1–2 (2003): 55–66.

Kieser, D.G., *System des Tellurismus oder thierischen Magnetismus: Ein Handbuch fur Naturforscher und Aerzte*, 2nd edition (Leipzig: F.L. Hervig, 1826).

Kirkpatrick, T. Percy, *The Book of the Rotunda Hospital: An Illustrated History of the Dublin Lying-In Hospital from Its Foundation in 1745 to the Present Time* (London: Bartholomew, 1913).

Kittler, Friedrich A., *Gramophone, Film, Typewriter* (Stanford: Stanford University Press, 1999).

Knapp, Steven and Michaels, Walter Benn, 'Against Theory', *Critical Inquiry*, 8.4 (1982): 723–42.

Knapp, Steven and Michaels, Walter Benn, 'A Reply to Our Critics', *Critical Inquiry*, 9.4 (1983): 790–800.

Komar, Kathleen L., 'Candide in Cyberspace: Electronic Texts and the Future of Comparative Literature', *Comparative Literature*, 59.3 (2007): vii–xviii.

Kottler, Malcolm Jay, 'Alfred Russel Wallace, the Origin of Man and Spiritualism', *Isis*, 65 (1974): 144–92.

Kraft, Hartmut, *Grenzgänge zwischen Kunst und Psychiatrie* (Cologne: Dumont Verlag, 1986).

Krauß, Friedrich, *Nothschrei eines Magnetisch-Vergifteten (1852) und Nothgedrungene Fortsetzung meines Nothschrei (1867): Selbstschilderungen eines Geisteskranken*, ed. H. Ahlenstiel and J.E. Meyer (Göttingen: Bayer-Leverkusen, 1967).

Kress, Jill M., *The Figure of Consciousness: William James, Henry James and Edith Wharton* (New York: Routledge, 2002).

Kurzweil, Ray, *The Age of Spiritual Machines: When Computers Exceed Human Intelligence* (New York: Viking, 1999).

Kurzweil, Ray, *Ray Kurzweil's Cybernetic Poet* (2001), at: www.kurzweilcyberart. com/poetry/rkcp_overview.php3.

Lacan, Jacques, *The Seminar I: Freud's Papers on Technique, 1953–54* (New York: Norton, 1988).

Landes, Joan B., 'The Anatomy of Artificial Life: An Eighteenth-Century Perspective', in Jessica Riskin, ed., *Genesis Redux: Essays in the History and Philosophy of Artificial Life* (Chicago: Chicago University Press, 2007): 96–116.

Latour, Bruno, *Pandora's Hope* (Cambridge, MA: Harvard University Press, 1999).

Law, Jules David, *The Rhetoric of Empiricism: Language and Perception from Locke to I. A. Richards* (Ithaca: Cornell University Press, 1993).

Leadbeater, Charles, *Living on Thin Air: The New Economy* (London: Penguin, 1999).

Leary, David E., *Metaphors in the History of Psychology* (Cambridge: Cambridge University Press, 1994).

Leatherdale, W.H., *The Role of Analogy, Model and Metaphor in Science* (Amsterdam: North Holland Publishing, 1974).

Lewes, George Henry, 'The Course of Modern Thought', *Fortnightly Review*, New Series, 21 (1877): 317–27.

Lewes, George Henry, *The Physiology of Common Life* (Leipzig: Bernard Tauchnitz, 1860).

Lewes, George Henry, *Problems in Life and Mind: The Foundation of a Creed* (London: Trübner, 1874).

Lewes, George Henry, *Problems in Life and Mind: Mind as a Function of the Organism* (London: Trübner, 1879).

Lewes, George Henry, 'Spinoza's Life and Works', *Westminster Review*, 39 (1843): 372–407.

Lewis, Matthew, *The Monk: A Romance* (Waterford: Printed for J. Saunders, 1796).

Lieske, Pam, 'William Smellie's Use of Obstetrical Machines and the Poor', *Studies in Eighteenth-Century Culture*, 29 (2000): 65–86.

Locke, John, *An Essay Concerning Human Understanding*, ed. Peter H. Nidditch (Oxford: Clarendon, 1975).

Locke, John, *An Essay Concerning Human Understanding*, ed. Alexander Campbell Fraser, 2 vols (New York: Dover, 1959).

Loeb, Lori, 'Beating the Flu: Orthodox and Commercial Responses to Influenza in Britain 1889–1919', *Social History of Medicine*, 18.2 (2005): 203–24.

Loren, Lewis A. and Dietrich, Eric, 'Merleau-Ponty, Embodied Cognition and the Problem of Intentionality', *Cybernetics and Systems: An International Journal*, 28 (1997): 345–58.

Luckhurst, Roger, *The Invention of Telepathy, 1879–1901* (Oxford: Oxford University Press, 2002).

Lucretius, Titus, *De Rerum Natura*, trans. Charles E. Bennett (Roslyn: Walter J. Black, 1946).

Lucretius, Titus, *De Rerum Natura*, 3rd edition, trans. H.A.J. Munro (Cambridge: Deighton Bell, 1886).

Mack, Douglas, 'Hogg and Angels', *Studies in Hogg and His World*, 15 (2004): 90–98.

Marchant, James (ed.), *Alfred Russel Wallace: Letters and Reminiscences*, 2 vols (London, New York, Toronto and Melbourne: Cassell and Company, 1916).

Margetts, Edward L., 'Trepanation of the Skull by the Medicine Men of Primitive Cultures, with Particular Reference to Present-day East-African Practice', in Donald R. Brothwell and A.T. Sandison, eds, *Diseases in Antiquity* (Springfield: Charles C. Thomas, 1967): 673–701.

Martin, Emily, 'The End of the Body', *American Ethnologist*, 19.1 (1992): 121–40.

Maxwell, James Clerk, 'Molecular Evolution', *Nature*, 8.205 (2 October, 1873): 473.

Mazlish, Bruce, 'The Man-Machine and Artificial Intelligence', in Ronald Chrisley, ed., *Artificial Intelligence: Critical Concepts* (London: Routledge, 2000): 134–60.

McCalman, Iain, *Darwin's Armada: How Four Voyagers to Australia Won the Battle for Evolution and Changed the World* (Melbourne: Penguin, 2009).

McHale, Brian, 'Poetry as Prosthesis', *Poetics Today*, 21.1 (2000): 3.

McKenzie, Morrell, 'Influenza', *Fortnightly Review*, 55 (1891): 877–86.

Meredith, George, *Diana of the Crossways: A Novel* (Detroit: Wayne State University Press, 2001).

Meredith, George, *The Ordeal of Richard Feverel* (London: Penguin, 1998).

Miller, Florence Fenwick, 'The Ladies Column', *Illustrated London News*, 96 (1890): 154–55.

Mitchell, W.J.T., *Against Theory: Literary Studies and the New Pragmatism* (Chicago: Chicago University Press, 1985).

Moholy-Nagy, Laszlo, 'A New Instrument of Vision', in Liz Wells, ed., *The Photography Reader* (London: Routledge, 2003): 92–96.

Moore, James R., 'Wallace's Malthusian Moment: The Common Context Revisited', in Bernard Lightman, ed., *Victorian Science in Context* (Chicago and London: University of Chicago Press, 1997): 290–311.

Moscucci, Ornella, *The Science of Woman: Gynaecology and Gender in England, 1800–1929* (Cambridge: Cambridge University Press, 1990).

Myers, Gerald E., *William James: His Life and Thought* (New Haven: Yale University Press, 1986).

Newton, Isaac, *Opticks or A Treatise of the Reflections, Refractions, Inflections and Colours of Light*, 4th edition (New York: Dover, 1952).

Niven, W.D. (ed.), *The Scientific Papers of James Clerk Maxwell* (Cambridge: Cambridge University Press, 1890).

North, Michael, *Camera-Works: Photography and the Twentieth-Century Word* (Oxford: Oxford University Press, 2005).

O'Halloran, Meiko, 'Treading the Borders of Fiction: Veracity, Identity and Corporeality in *The Three Perils*', *Studies in Hogg and His World*, 12 (2001): 40–55.

Paley, William, *Natural Theology or, Evidences of the Existence and Attributes of the Deity, Collected from the Appearances of Nature* (London: Fauldner, 1810).

Paterson, Samuel, *A Catalogue of the Entire and Inestimable Apparatus for Lectures in Midwifery, Contrived with Consummate Judgment, and Executed with Infinite Labour, by the Late Ingenious Dr. William Smellie, Deceased: Consisting of a Variety of Anatomical Preparations, Illustrating the Theory of Midwifery, the Original Drawings by Rymsdyk, from which his Engravings were Made, his Exquisite*

Artificial Machines, in Imitation of the Living Subjects, his Collection of Obstetrical Instruments, English and Foreign (London: Royal College of Obstetricians and Gynaecologists, 1770).

Patterson, David K., *Pandemic Influenza, 1700–1900: A Study in Historical Epidemiology* (Totowa: Rowman and Littlefield, 1986).

Payn, James, 'Our Notebook', *Illustrated London News*, 96 (1890): 98.

Pendergrast, Mark, *Mirror | Mirror: A History of the Human Love Affair with Reflection* (New York: Basic Books, 2003).

Penniman, Charles, 'Maillardet's Automaton', *The Franklin Institute*, at: www.fi.edu/learn/scitech/automaton/automaton.php?cts=instrumentation.

Perceval, John, *A Narrative of the Treatment Experienced by a Gentleman, During a State of Mental Derangement: Designed to Explain the Causes and the Nature of Insanity, and to Expose the Injudicious Conduct Pursued Towards Many Sufferers Under That Calamity* (London: Effingham Wilson, 1838).

Perceval, John, *Perceval's Narrative: A Patient's Account of His Psychosis 1830–1832* (Stanford: Stanford University Press, 1961).

Pettigrew, Thomas Joseph, *A History of Egyptian Mummies* (London: Longmans, 1834).

Pettitt, Clare, 'On Stuff', *19: Interdisciplinary Studies in the Long Nineteenth Century*, 6 (2008): 1–12.

Porter, Roy, 'Introduction', in *Illustrations of Madness*, by John Haslam (London and New York: Routledge, 1988): xi–lxiv.

Prentice, Rachel, 'The Visible Human', in Sherry Turkle, ed., *The Inner History of Devices* (Cambridge, MA and London: The MIT Press, 2008).

Pulham, Patricia, 'The Eroticism of Artificial Flesh in Villiers de L'Isle Adam's *L'Eve Future*', in 'Minds, Bodies, Machines', Deirdre Coleman and Hilary Fraser eds, *19: Interdisciplinary Studies in the Long Nineteenth Century*, 7 (2008).

Radcliffe, Ann, *The Mysteries of Udolpho* (Oxford: Oxford University Press, 1988).

Radcliffe, Ann, 'On the Supernatural in Poetry', in E.J. Clery and Robert Miles, eds, *Gothic Documents: A Sourcebook 1700–1820* (Manchester and New York: Manchester University Press, 2000): 163–72.

Raichle, Marcus E. and Mintun, Mark A., 'Brain Work and Brain Imaging', *Annual Review of Neuroscience*, 29 (2006): 449–76.

Reid, Thomas, *An Inquiry into The Human Mind, on the Principles of Common Sense, with an Account of the Life and Writings of the Author* (Edinburgh: Printed for A. Millar, London, and A. Kincaid & J. Bell, Edinburgh, 1764).

Richards, Thomas, *The Commodity Culture of Victorian England: Advertising and Spectacle, 1851–1914* (London: Verso, 1991).

Rieger, Stefan, 'Psychopaths Electrified: Die Wahnwege des Wissens im *Notschrei eines Magnetisch-Vergifteten*', in Torsten Hahn, Jutta Person and Nicolas Pethes, eds, *Grenzgänge zwischen Wahn und Wissen: zur Koevolution von Experiment und Paranoia, 1850–1910* (Frankfurt and New York: Campus, 2002): 151–72.

Riskin, Jessica, 'The Defecating Duck, or, the Ambiguous Origins of Artificial Life', *Critical Inquiry*, 29.4 (2003): 599–633.

Riskin, Jessica, 'Eighteenth-Century Wetware', *Representations*, 83 (2003): 97–125.

Riskin, Jessica, *Science in the Age of Sensibility: The Sentimental Empiricists of the French Enlightenment* (Chicago: University of Chicago Press, 2002).

Romanes, George John, *Mind and Motion and Monism* (London: Longmans, Green, 1895).

Rylance, Rick, *Victorian Psychology and British Culture, 1850–1880* (Oxford: Oxford University Press, 2000).

Salisbury, Laura, 'Sounds of Silence: Aphasiology and the Subject of Modernity', in Laura Salisbury and Andrew Shail, eds, *Neurology and Modernity* (Basingstoke: Palgrave Macmillan, 2010).

Salisbury, Laura, '"What Is the Word": Beckett's Aphasic Modernism', *Journal of Beckett Studies*, 17 (2008): 78–126.

Schaffer, Simon, 'Enlightened Automata', in William Clark, Jan Golinski and Simon Schaffer, eds, *The Sciences in Enlightened Europe* (Chicago: University of Chicago Press, 1999): 126–65.

Schofield, Alfred, 'A Few Notes on Influenza', *Leisure Hour*, 40 (1891): 570.

Seltzer, Mark, *Bodies and Machines* (New York and London: Routledge, 1992).

Serres, Michel, *The Birth of Physics*, trans. Jack Hawkes (Manchester: Clinamen Press, 2000).

Shermer, Michael, *In Darwin's Shadow: The Life and Science of Alfred Russel Wallace* (Oxford: Oxford University Press, 2004).

Sherry, Vincent, *The Great War and the Language of Modernism* (Oxford: Oxford University Press, 2003).

Sicker, Philip, *Love and the Quest for Identity in the Fiction of Henry James* (Princeton: Princeton University Press, 1980).

Sidney, Sir Philip, *An Apology for Poetry: or, The Defence of Poesy*, ed. Geoffrey Shepherd (Manchester: Manchester University Press, 2002).

Siegert, Bernhard, 'Gehörgänge ins Jenseits: Der telephonistische Entzug des Ohres', in Torsten Hahn, Jutta Person and Nicolas Pethes, eds, *Grenzgänge zwischen Wahn und Wissen: zur Koevolution von Experiment und Paranoia, 1850–1910* (Frankfurt and New York: Campus, 2002): 173–92.

Simpson, A.R., 'History of Chair of Midwifery and the Diseases of Women and Children in the University of Edinburgh', *Edinburgh Medical Journal*, 29 (1998): 481–98.

Sleigh, Charlotte, 'Life, Death and Galvanism', *Studies in History and Philosophy of Biological and Biomedical Sciences*, 29 (1998): 219–48.

Slotten, Ross A., *The Heretic in Darwin's Court: The Life of Alfred Russel Wallace* (New York: Columbia University Press, 2004).

Smellie, William, *A Course of Lectures upon Midwifery, wherein the Theory and Practice of that Art are Explain'd in the Clearest Manner. More particularly, the Structure of the Pelvis and Uterus. Of the Foetus in Utero, and after Parturition. The Management of Child-Bearing Women, during Pregnancy, in time of Labour, and after Delivery. The Manner of Delivering Women, in all the Variety of Natural, Difficult, and Preternatural Labours, Perform'd on Different Machines made in Imitation of Real Women and Children* (London, 1745, Wellcome Library MS 4630).

Smellie, William, *A Sett of Anatomical Tables, With Explanations, and an Abridgment, of the Practice of Midwifery, With a View to Illustrate a Treatise On that Subject, and Collection of Cases* (London, 1754).

Smellie, William, *Smellie's Treatise on the Theory and Practice of Midwifery* (London: The New Syndenham Society, 1876).

Smith, Charles H., 'Spritualism, and Beyond: "Change," or "No Change"?', in Charles H. Smith and George Beccaloni, eds, *Natural Selection & Beyond: The Intellectual Legacy of Alfred Russel Wallace* (Oxford and New York: Oxford University Press, 2008): 391–423.

Smith, F.B., 'The Russian Influenza in the United Kingdom, 1889–1894', *Social History of Medicine*, 8.1 (1995): 55–73.

Smith, Roger, 'Alfred Russel Wallace: Philosophy of Nature and Man', *British Journal for the History of Science*, 6 (1972): 177–99.

Sobchack, Vivian, *Carnal Thoughts: Embodiment and Moving Image Culture* (Berkeley: University of California Press, 2004).

Sommers, Carl, 'By the Way; Inspiration or Computation', *New York Times* (28 November 1999) at: www.nytimes.com/1999/11/28/nyregion/by-the-way-inspiration-or-computation.html?n=Top%2FReference%2FTimes%20Topics%2FSubjects%2FW%2FWriting%20and%20Writers.

Spafford, Eugene H., 'Computer Viruses as Artificial Life', *Journal of Artificial Life*, 1.3 (1994): 249–65.

Spencer, Herbert, *First Principles* (London: Williams and Norgate, 1862).

Spencer, Herbert, *Principles of Biology*, 2 vols (London: Williams and Norgate, 1864–67).

Spencer, Herbert, *Principles of Psychology* (London: Longman, Brown, Green and Longmans, 1855).

Spencer, Herbert, *Principles of Psychology*, 2nd edition, 2 vols (London: Williams and Norgate, 1870–72).

Stich, Stephen P. (ed.), *Innate Ideas* (Berkeley: University of California Press, 1975).

Sully, James, 'The Aesthetics of Human Character', *Fortnightly Review*, New Series, 9 (1871): 505–20.

Sully, James, *My Life and Friends: A Psychologist's Memories* (London: Fisher Unwin, 1918).

Sully, James, *Outlines of Psychology: with Special Reference to Education*, 2nd edition (London: Longman's Green, 1885).

Sully, James, *Sensation and Intuition: Studies in Psychology and Ethics* (London: Henry S. King, 1874).

Tausk, Viktor, 'On the Origin of the "Influencing Machine" in Schizophrenia', in *Sexuality, War and Schizophrenia: Collected Psychoanalytic Papers*, ed. Paul Roazen, trans. Dorian Feigenbaum (New Brunswick: Transaction Publishers, 1991): 185–219.

Thatcher, John, *A Letter to the Lord Provost and Patrons of the University of Edinburgh, On the Proposed New Regulations Respecting the Study of Midwifery* (Edinburgh: Carfrae, 1825).

Thurschwell, Pamela, *Literature, Technology and Magical Thinking, 1880–1920* (Cambridge: Cambridge University Press, 2001).

Turner, Frank M., *Contesting Cultural Authority: Essays in Victorian Intellectual Life* (Cambridge: Cambridge University Press, 1993).

Tyndall, John, *Fragments of Science* (New York: D. Appleton and Company, 1892).

Ure, Andrew, 'An Account of Some Experiments Made on the Body of a Criminal Immediately after Execution, with Physiological and Practional Observations', *Journal of Science and the Arts*, 6 (1819): 283–94.

Van Loon, Joost, 'A Contagious Living Fluid: Objectivity and Assemblage in the History of Virology', *Theory, Culture and Society*, 19.5–6 (2002): 107–24.

Virchow, Rudolf, 'On the Mechanistic Interpretation of Life', *Disease, Life and Man: Selected Essays by Rudolf Virchow*, trans. Lelland J. Rather (Stanford: Stanford University Press, 1958): 102–19.

von Helmholtz, Hermann, 'On the Conservation of Force', in *Science and Culture: Popular and Philosophical Essays*, trans. David Cahan (Chicago: University of Chicago Press, 1995): 96–126.

Waldby, Catherine, *The Visible Human Project: Informatic Bodies and Posthuman Medicine* (London and New York: Routledge, 2000).

Wallace, Alfred Russel, *My Life. A Record of Events and Opinions* (London: Chapman & Hall, 1908; facsimile copy, Elibron Classics, 2005).

Wallace, Alfred Russel, *My Life: A Record of Events and Opinions*, 2 vols (London: Bell, 1905).

Wallace, Alfred Russel, *Contributions to the Theory of Natural Selection* (London, New York: Macmillan, 1870).

Wallace, Alfred Russel, 'The Origin of Human Races and the Antiquity of Man Deduced from the Theory of Natural Selection', [a paper read at the ASL meeting of 1 March 1864], *Journal of the Anthropological Society of London*, 2 (1864): clviii–clxx.

Wallace, Alfred Russel, *The Wonderful Century* (London and New York: Swan, Sonnenschein, 1899).

Waller, P.J., *Town, City and Nation: England, 1850–1914* (Oxford: Oxford University Press, 1983).

Walpole, Horace, 'Preface to the First Edition', in *The Castle of Otranto: A Gothic Story*, ed. W.S. Lewis (Oxford: Oxford University Press, 1986): 3–6.

Warner, Marina, *Phantasmagoria: Spirit Visions, Metaphors and Visual Media into the Twenty-First Century* (Oxford: Oxford University Press, 2006).

Warwick, Kevin, *I, Cyborg* (Urbana and Chicago: University of Illinois Press, 2002).

Wernicke, Carl, 'The Motor Speech Path and the Relation of Aphasia to Anarthria', in Gertrude H. Eggert, ed., *Wernicke's Works on Aphasia: A Sourcebook and Review* (1884; repr., The Hague: Mouten, 1977).

West, David, *Horace Odes I, Carpe Diem: Text, Translation and Commentary* (Oxford: Clarendon Press, 1995).

Williams-Ellis, Amabel, *Darwin's Moon: A Biography of Alfred Russel Wallace* (London: Blackie, 1966).

Wilson, Andrew, 'Science Jottings: Our Monthly Look-Around', *Illustrated London News*, 96 (1890): 146.

Wilson, Elizabeth, 'Imaginable Computers: Affects and Intelligence in Alan Turing', in Darren Tofts, Annemarie Jonson and Alessio Cavallaro, eds, *Prefiguring Cyberculture: An Intellectual History* (Cambridge MA: MIT Press, 2004): 38–51.

Wimsatt, William K. and Beardsley, Monroe C., 'The Intentional Fallacy', in William K. Wimsatt, ed., *The Verbal Icon: Studies in the Meaning of Poetry* (Lexington: University of Kentucky Press, 1954): 3–18.

Winter, Alison, *Mesmerized: Powers of Mind in Victorian Britain* (Chicago and London: Chicago University Press, 1998).

Artificial Machines, in Imitation of the Living Subjects, his Collection of Obstetrical Instruments, English and Foreign (London: Royal College of Obstetricians and Gynaecologists, 1770).

Patterson, David K., *Pandemic Influenza, 1700–1900: A Study in Historical Epidemiology* (Totowa: Rowman and Littlefield, 1986).

Payn, James, 'Our Notebook', *Illustrated London News*, 96 (1890): 98.

Pendergrast, Mark, *Mirror | Mirror: A History of the Human Love Affair with Reflection* (New York: Basic Books, 2003).

Penniman, Charles, 'Maillardet's Automaton', *The Franklin Institute*, at: www.fi.edu/learn/scitech/automaton/automaton.php?cts=instrumentation.

Perceval, John, *A Narrative of the Treatment Experienced by a Gentleman, During a State of Mental Derangement: Designed to Explain the Causes and the Nature of Insanity, and to Expose the Injudicious Conduct Pursued Towards Many Sufferers Under That Calamity* (London: Effingham Wilson, 1838).

Perceval, John, *Perceval's Narrative: A Patient's Account of His Psychosis 1830–1832* (Stanford: Stanford University Press, 1961).

Pettigrew, Thomas Joseph, *A History of Egyptian Mummies* (London: Longmans, 1834).

Pettitt, Clare, 'On Stuff', *19: Interdisciplinary Studies in the Long Nineteenth Century*, 6 (2008): 1–12.

Porter, Roy, 'Introduction', in *Illustrations of Madness*, by John Haslam (London and New York: Routledge, 1988): xi–lxiv.

Prentice, Rachel, 'The Visible Human', in Sherry Turkle, ed., *The Inner History of Devices* (Cambridge, MA and London: The MIT Press, 2008).

Pulham, Patricia, 'The Eroticism of Artificial Flesh in Villiers de L'Isle Adam's *L'Eve Future*', in 'Minds, Bodies, Machines', Deirdre Coleman and Hilary Fraser eds, *19: Interdisciplinary Studies in the Long Nineteenth Century*, 7 (2008).

Radcliffe, Ann, *The Mysteries of Udolpho* (Oxford: Oxford University Press, 1988).

Radcliffe, Ann, 'On the Supernatural in Poetry', in E.J. Clery and Robert Miles, eds, *Gothic Documents: A Sourcebook 1700–1820* (Manchester and New York: Manchester University Press, 2000): 163–72.

Raichle, Marcus E. and Mintun, Mark A., 'Brain Work and Brain Imaging', *Annual Review of Neuroscience*, 29 (2006): 449–76.

Reid, Thomas, *An Inquiry into The Human Mind, on the Principles of Common Sense, with an Account of the Life and Writings of the Author* (Edinburgh: Printed for A. Millar, London, and A. Kincaid & J. Bell, Edinburgh, 1764).

Richards, Thomas, *The Commodity Culture of Victorian England: Advertising and Spectacle, 1851–1914* (London: Verso, 1991).

Rieger, Stefan, 'Psychopaths Electrified: Die Wahnwege des Wissens im *Notschrei eines Magnetisch-Vergifteten*', in Torsten Hahn, Jutta Person and Nicolas Pethes, eds, *Grenzgänge zwischen Wahn und Wissen: zur Koevolution von Experiment und Paranoia, 1850–1910* (Frankfurt and New York: Campus, 2002): 151–72.

Riskin, Jessica, 'The Defecating Duck, or, the Ambiguous Origins of Artificial Life', *Critical Inquiry*, 29.4 (2003): 599–633.

Riskin, Jessica, 'Eighteenth-Century Wetware', *Representations*, 83 (2003): 97–125.

Riskin, Jessica, *Science in the Age of Sensibility: The Sentimental Empiricists of the French Enlightenment* (Chicago: University of Chicago Press, 2002).

Romanes, George John, *Mind and Motion and Monism* (London: Longmans, Green, 1895).

Rylance, Rick, *Victorian Psychology and British Culture, 1850–1880* (Oxford: Oxford University Press, 2000).

Salisbury, Laura, 'Sounds of Silence: Aphasiology and the Subject of Modernity', in Laura Salisbury and Andrew Shail, eds, *Neurology and Modernity* (Basingstoke: Palgrave Macmillan, 2010).

Salisbury, Laura, '"What Is the Word": Beckett's Aphasic Modernism', *Journal of Beckett Studies*, 17 (2008): 78–126.

Schaffer, Simon, 'Enlightened Automata', in William Clark, Jan Golinski and Simon Schaffer, eds, *The Sciences in Enlightened Europe* (Chicago: University of Chicago Press, 1999): 126–65.

Schofield, Alfred, 'A Few Notes on Influenza', *Leisure Hour*, 40 (1891): 570.

Seltzer, Mark, *Bodies and Machines* (New York and London: Routledge, 1992).

Serres, Michel, *The Birth of Physics*, trans. Jack Hawkes (Manchester: Clinamen Press, 2000).

Shermer, Michael, *In Darwin's Shadow: The Life and Science of Alfred Russel Wallace* (Oxford: Oxford University Press, 2004).

Sherry, Vincent, *The Great War and the Language of Modernism* (Oxford: Oxford University Press, 2003).

Sicker, Philip, *Love and the Quest for Identity in the Fiction of Henry James* (Princeton: Princeton University Press, 1980).

Sidney, Sir Philip, *An Apology for Poetry: or, The Defence of Poesy*, ed. Geoffrey Shepherd (Manchester: Manchester University Press, 2002).

Siegert, Bernhard, 'Gehörgänge ins Jenseits: Der telephonistische Entzug des Ohres', in Torsten Hahn, Jutta Person and Nicolas Pethes, eds, *Grenzgänge zwischen Wahn und Wissen: zur Koevolution von Experiment und Paranoia, 1850–1910* (Frankfurt and New York: Campus, 2002): 173–92.

Simpson, A.R., 'History of Chair of Midwifery and the Diseases of Women and Children in the University of Edinburgh', *Edinburgh Medical Journal*, 29 (1998): 481–98.

Sleigh, Charlotte, 'Life, Death and Galvanism', *Studies in History and Philosophy of Biological and Biomedical Sciences*, 29 (1998): 219–48.

Slotten, Ross A., *The Heretic in Darwin's Court: The Life of Alfred Russel Wallace* (New York: Columbia University Press, 2004).

Smellie, William, *A Course of Lectures upon Midwifery, wherein the Theory and Practice of that Art are Explain'd in the Clearest Manner. More particularly, the Structure of the Pelvis and Uterus. Of the Foetus in Utero, and after Parturition. The Management of Child-Bearing Women, during Pregnancy, in time of Labour, and after Delivery. The Manner of Delivering Women, in all the Variety of Natural, Difficult, and Preternatural Labours, Perform'd on Different Machines made in Imitation of Real Women and Children* (London, 1745, Wellcome Library MS 4630).

Smellie, William, *A Sett of Anatomical Tables, With Explanations, and an Abridgment, of the Practice of Midwifery, With a View to Illustrate a Treatise On that Subject, and Collection of Cases* (London, 1754).

Smellie, William, *Smellie's Treatise on the Theory and Practice of Midwifery* (London: The New Syndenham Society, 1876).

Index

Wolfe, Cary, *What is Posthumanism?* (Minneapolis and London: University of Minnesota Press, 2010).

Wood, Gaby, *Edison's Eve: A Magical History of the Quest for Mechanical Life* (New York: Knopf, 2002).

Wordsworth, William, *The Oxford Authors: William Wordsworth*, ed. Stephen Gill (Oxford: Oxford University Press, 1984).

Wunnicke, Christine, '"Auserwählt zum Aufbruch": Der bürgerliche Wahsninn des Friedrich Krauß', in Torsten Hahn, Jutta Person and Nicolas Pethes, eds, *Grenzgänge zwischen Wahn und Wissen: zur Koevolution von Experiment und Paranoia, 1850—1910* (Frankfurt and New York: Campus, 2002): 110–24.

Young, Robert M., *Mind, Brain and Adaptation* (Oxford: Oxford University Press, 1980).